Inventing TV News

Live and Local in Los Angeles

Terry Anzur

Copyright © 2022 Theresa Marie Anzur

All rights reserved.

Introduction

The bare light bulb came on automatically as I stepped through the doorway of the musty basement. I made my way across the uneven cement floor, avoiding a large, dead cockroach. Dust-covered boxes partially blocked my path to the rusty file cabinets against the back wall. Pulling open a drawer, I found yellowed folders of crumbling newspaper clippings and office memos. Glossy black-and-white photographs were stuck together in useless clumps. Judging from the dampness and the mold, the place had been flooded at one time and then abandoned.

Above me, the busy KTLA TV newsroom hummed with the routine of reporting the day's top stories. I was the co-anchor of the station's 10 o'clock newscast. At the same time, I was holding down a second job teaching broadcast journalism at the University of Southern California. I was certainly the first and probably the last tenure-track faculty member to appear on billboards all over Los Angeles as one of the 'authentic LA' faces of News at Ten. But that fun fact wasn't going to help me win tenure at the prestigious Annenberg School of Communication and Journalism.

I had been advised that broadcasting nightly on an award-winning, top-rated news program didn't count as 'publishing in my field.' In 1997, the school had not yet established a professional track for experienced journalists lacking a traditional PhD. I was going to have to earn tenure the old-fashioned way: publish or perish. KTLA's management had given me permission to look through the basement archives in my spare time. I quickly realized I had access to a treasure trove of primary source documents on the early development of local TV news in Los Angeles. For the next three years, I did research by reading the files and interviewing both retired and active KTLA colleagues. They kindly shared their personal recollections.

Local television news has been notoriously unkind to its past. Some early TV shows were preserved on film kinescopes. But local news programs, aired primarily as a public service obligation and of no value in the syndication market, usually were not considered worth saving. After videotape was invented in the late 1950s, the bulky two-inch reels and three-quarter-inch cassettes presented a costly storage problem. News videotapes were recorded over or thrown away as stations changed hands. News film was recycled for the silver. Much of the early history of local TV news survives only in the memories of the people who launched this industry and the paper trail they left behind. For the broadcast journalism historian, this means a race not only against the clock but also against the short-term memory of a business that values the ratings for the next big story over the lessons that could be learned from the past.

Broadcast historians have focused on the achievements of national news divisions at the three major US broadcasting networks, mostly overlooking significant developments at the local level. As my USC colleague and academic mentor Joe

Saltzman has noted, "the Great Television Networks—those who work for them, watch them and criticize them—consider local television much as a dog considers fleas: an annoying, insignificant, brutally tiresome fact of life."[1]

Some scholars present local television news primarily as the competitor that killed the evening newspaper, ignored vital community issues, valued 'happy talk' over substance, and reduced political discourse to something less than a 15-second sound bite. Such criticism fails to credit local television news for what it does well. For live coverage of breaking news, and especially when a community is coping with natural disaster or man-made calamity, local TV news has no equal. The networks can't be everywhere; they initially rely on their local station affiliates when news breaks. In virtually every American city, there is fierce rivalry among local TV news stations. They remain intensely competitive even when there is only one surviving newspaper in the community they serve.

The battle between independent stations KTLA and KTTV in the early days encouraged technological innovation. Led by the visionary Klaus Landsberg, KTLA pioneered live, on-the-scene reporting from the ground and from the air, even bringing the atomic bomb into America's living rooms when the national networks said it couldn't be done. These achievements have been well documented in books by legendary KTLA reporter Stan Chambers and journalist Evelyn DeWolfe, who was married to Landsberg during KTLA's formative years. I wanted to add to this KTLA-centered version of local TV news history by placing it in a wider context.

I read contemporary newspaper accounts of the same events, noting what print journalists and critics thought of the TV coverage by all of the competing stations in Los Angeles. I also viewed the actual programs that were available in the archives at UCLA and the Museum of Broadcasting. When the programs were not preserved it was necessary to reconstruct the coverage from newspaper and eyewitness accounts, as well as interviews with actual participants. There is no 'invented' dialog. Any quotations are from a recording or a published source, or the exact words as remembered by someone who was there.

Let's open those rusty file cabinets and revisit the invention of live and local television news in Los Angeles:

> **Everyone's Child, 1949:** The race to rescue three-year-old Kathy Fiscus from an abandoned well pipe was the first unscheduled breaking news event to be covered by two competing TV stations with live remote capability. The broadcasts also marked the first time a family suffered the loss of a loved one in full view of a television audience. The 27 hours of extended live coverage from the scene made celebrities out of first responders and the 'sandhogs' who risked their lives in the attempt to save a little girl. It showed consumers that TV was more than just a gadget: live television could transmit a compelling story into their homes as it unfolded.
>
> **Covering Crime, 1951:** The investigation into the murder of eight-year-old Patty Jean Hull, kidnapped

from a movie theater by a known sex offender, raised ethical questions about sensational crime coverage. While KTLA hesitated to bring the lurid details into viewers' living rooms, KTTV offered a more aggressive approach. The extended live coverage foreshadowed the development of crime-centered local newscasts and the motto, "If it bleeds, it leads."

The Cold War Hits Home, 1952: KTLA's Klaus Landsberg took on the challenge of televising an atomic bomb detonation, risking everything to overcome technical and logistical obstacles. One local station brought live pictures of the Cold War into homes across America when the networks insisted it was impossible.

Pictures From the Air, 1958: The top-secret project that launched the world's first TV news helicopter, capable of transmitting live pictures from the air. Aerial coverage gave KTLA an advantage over its competitors in reporting the Bel Air fire and the Baldwin Hills dam collapse. No other TV station in America would match KTLA's technological feat until 1974, when the Telecopter was sold to a rival station in Los Angeles.

The Rise of the Celebrity Anchorman, 1970s: Two newscasters in Los Angeles, George Putnam and Jerry Dunphy, were the role models for the iconic character of Ted Baxter on the Mary Tyler Moore

Show. Local TV news presenters evolved into highly paid celebrities, with appearance and personality often valued over journalistic ability.

You, the reader, are invited to journey back in time to meet the fascinating heroes and rogues who invented live and local TV news in Los Angeles.

Chapter One:
"Everyone's Child"

This is about the birth of live, on-the-scene television in Los Angeles, about how the car chases and the reporters in front of "live" courthouses began. It is about one of the most influential figures in Los Angeles television—a little girl who in all likelihood never even saw a television program, and then became one.[2]

—Patt Morrison, Los Angeles Times Magazine, "The Little Girl Who Changed Television Forever"

Alice Fiscus looked out her kitchen window and counted her blessings. It was Friday, April 8, 1949. She was a typical suburban mom on a sunny, late afternoon in Southern California. With so many returning servicemen and their sweethearts now married and raising families in these post-World War II years, good houses were selling for $10,000. The Fiscus family had endured a series of moves as eager buyers snapped up the few rental houses available. Soon, they would have their own home on a newly developed parcel of land in San Marino, an upper middle-class community 11 miles northeast of downtown of Los Angeles. The site for the new house was just one block away from the yellow, wood-frame farmhouse they were renting at 2590 Robles Avenue. The rental property was dotted with fruit trees, reminders of the San Gabriel Valley's agricultural past. But it had been years since there were crops in the vacant field and horses in the barn. Southern Californians depended more and more on their cars to get around and plenty of ex-GI's could afford $1,700 for a shiny new automobile. A new elementary school was being built at the opposite edge of the field to educate the children of the post-war baby boom.

Alice was busy in the kitchen. She was also enjoying the company of her sister, Jeanette Lyon, who had arrived that morning by train from Chula Vista, their hometown. From the window Alice could see her daughters laughing and running down a grassy slope in the vacant lot behind the house with their two cousins, Stanley, age 9 and Gus, age 5. Nine-year-old Barbara sprinted ahead of the boys and younger sister Kathy, who bounced over the grass in her pink party dress. Blonde-haired and blue-eyed with a wide, gap-toothed grin, three-year-old Kathy tried hard to keep up with the others. It was Barbara's responsibility, as the eldest, to make sure they did not go past the play-area boundary her parents had set at the wire fence. When she reached "the limits," Barbara whirled around and headed back to the house, the cousins following her lead. For a moment, Alice turned her attention to what she would serve the four children for dinner. When she glanced outside again, they were halfway across the lot.

And there were only three of them.

Inventing TV News

Alice raced out of the kitchen, banging the door to the screened back porch as she charged out to the field. Jeanette followed closely behind, calling to the children.

"Where's Kathy?"

Barbara and the boys answered with puzzled looks. Wasn't Kathy right behind them? They looked around, then fanned out across the field, calling the little girl's name. In disbelief, Alice scanned the open area, about the size of a city block, between the house and the elementary school grounds. Even more worried now, she jumped in the car and drove around the block. There was no trace of Kathy. It was almost 5 p.m.

Jeanette and the children were still searching in the field. Alice grabbed the phone to call the police and her husband's office. Minutes later, San Marino police officers and firefighters arrived and joined the search. About 200 yards from the house, Kathy's kindergarten-age cousin stopped. Gus thought he heard something, like the whimper of a small animal. Was it his imagination? The muffled cries appeared to be coming from under the ground.

"There!"

The boy yelled and pointed to a hole, nearly overgrown by grass and weeds. Alice hurried over to where Gus was standing. As she threw herself onto the ground, she could hear her baby sobbing. Horrified, she realized the opening was no wider than a frying pan.

"Can you hear me, Kathy?" Alice called, peering into the pit.

"Yes..." The fearful little voice rose plaintively from the blackness below. Alice reached in and flailed her arm in the shaft, touching only the rusty metal sides of the hole. There was no way to tell how deep it was. She kept talking to Kathy and tried to make her voice sound as comforting as a hug. "Mommy's here, be brave."

Alice was startled by the touch of a hand on her shoulder. She looked up to see the concerned face of her husband.

"It's a miracle you found her," said Dave Fiscus, knowing that it might take a second miracle to get Kathy out of the pipe, only 14 inches wide. It appeared to be an abandoned well. He had walked this field many times and never knew it was there. Kneeling, he called softly to Kathy. "We'll get you out, honey. Don't be afraid."

Both parents saw the irony in their situation. Dave was Division Manager of the California Water and Telephone Company. Later, he would confirm that his employer dug—and later abandoned—the well in which his daughter was now imprisoned. He was just back from a trip to Sacramento, where he had unsuccessfully lobbied the state legislature for a bill to cap the abandoned wells and excavations that pockmarked the California landscape of former farms and oil fields.

Kathy's parents were well-known volunteers in the Red Cross and other community organizations. Dave had spent the war years in what was considered a crucial engineering job on the home front and knew most of the contractors in the area. Now, as he made a silent vow to do everything possible to rescue Kathy, he began calling friends in the construction business for help. Dave and Alice Fiscus were about to find out just how many friends they had.[3]

Inventing TV News

***

Los Angeles Times newsman Bill Johnston was interviewing an emergency room doctor who had a police radio. Strident voices crackled over the static, something about the fire department sending ropes. The cub reporter's gut told him it might be news. Maybe a person was caught on a telephone pole. It would make a decent picture, he thought. Johnston was a 'combination man,' covering the San Gabriel Valley beat with a notebook and a 1910-vintage speed-graphic camera. He headed for the press room of the Pasadena police department.

"Talk to the fire department," the dispatcher said. So, Johnston went to the fire department. Nothing going on, they claimed. But the police radio was still squawking: "Santa Anita and Robles..." Johnston phoned the Los Angeles County sheriff's office in Altadena. "Why don't you try San Marino?" he was told. San Marino police instructed him to call back later.

Johnston slammed down the phone. He hated getting the runaround. He dialed San Marino again. "Okay," a cop said, "there's a little girl who fell down a pipe."

The reporter called his night city editor to say that he was going to check it out.

Johnston was the first newsman to arrive at the Fiscus home. Days later he would write to his parents, "I saw a few cars parked and housewives standing in front of their homes, aprons on, looking in the direction of the field."

Police officers attempted to block the reporter's path to the fire trucks parked on the sloping lot. At 5:20 p.m. Johnston joined a small group of people next to the hole. From below, he could hear a child's muffled screams.

"I have never heard a more heart-breaking sound," he wrote to his parents. "The father kept moving toward and away from the hole, standing up, sitting down, lying flat with his head in the hole, standing up, walking away, rubbing his hands and his head, his face suddenly contorting in an effort to keep his emotion from overflowing."

Johnston took a closer look at the father and recognized Dave Fiscus as a friend of the publisher of the San Marino Tribune. The young newsman supplemented his income by selling advertising for the San Marino paper and had seen Dave with the publisher several times at the nearby Santa Anita racetrack.

"You know me, don't you, Dave?" Johnston asked. It took a moment for Kathy's father to reply. He was relieved to see a vaguely familiar face.

"Yes, I know you, Bill." Dave talked about what had happened to his younger daughter and Johnston scribbled notes. There was just enough light left to get some good pictures of the scene. Johnston filed his story and was told it would run as a banner headline in the early editions. It wouldn't be an exclusive for long.[4]

***

"It was frantic but well organized," Alice would later recall. Emergency crews and volunteers converged on the field. A hose was lowered into the hole and Pasadena firefighters took turns hand-cranking what appeared to be an oversized coffee

Inventing TV News

grinder. It was the pump that would provide fresh air for the trapped child to breathe. St. Luke's Hospital in nearby Pasadena, where Kathryn Ann Fiscus was born on August 21, 1945, now stationed an ambulance near the scene of the accident to rush the child to the emergency room the moment she emerged from the pit. Kathy's pediatrician joined her parents next to the hole. Dr. Robert J. McCullock heard the "brave little voice... like a child shut in a dark closet. There wasn't pain in it, just bewilderment."[5]

Firemen dangled a rope into the pipe as the little girl's parents shouted encouragement.

"Will you try to grab hold of the rope, Kathy?"

"I am, I AM."

It was 5:45 p.m. Rescuers studied the limp lifeline for any sign of a tug from a tiny hand. Kathy tried once to grasp the rope but couldn't hold on as firefighters attempted to pull her to safety. The adults gathered around the opening had no way to determine the child's position in the hole. They could still hear her crying but she was no longer tugging at the rescue line. Maybe she couldn't even see it. What if she had landed face down?

Local contractors brought pipes, pumps and digging equipment. Police sirens wailed as squad cars escorted heavy machinery to the scene. Members of the San Marino Canteen Corps set up a Red Cross area with food, drinks and a first aid station for the workers planning Kathy's rescue.

Already, the child's cries were becoming less frequent. Alice, straining to listen for the tiniest sound, kept up her own spirits by recalling how happily this day had started. Kathy had snuggled in her mother's arms as they waited at the train station for her cousins to arrive. Alice was still recovering from recent surgery and it was the first time in three months that she had been strong enough to pick up the active three-year-old.

Now, beside the hole, she cherished the memory of hugging her little girl and prayed that she would soon hold her again.

Alice and Dave took turns speaking to their daughter in her dark, rusty prison. "We could let her know we were still there," Alice would later recall. "Our hopes and prayers were always high for a successful rescue. We NEVER gave up."

Police put out a casting call to movie studios and Hollywood agents for anyone who might fit into the 14-inch hole. Among those who volunteered were child actors, jockeys from the nearby racetrack, little people from the Clyde Beatty Circus and even the boy from the "Call for Philip Morris" cigarette advertisements. However, rescuers vetoed this plan, fearing that anyone crawling into the well also might be trapped—or cut to ribbons by the badly corroded surface of the metal casing. By 6 p.m., they decided to dig a parallel shaft. They estimated that the little girl was about 100 feet down, trapped beyond a bend in the pipe.

As excavation began and the noise from the machines got louder, Kathy's parents could no longer hear their child's cries. Someone suggested that Alice sit down and rest. With her dark hair pulled straight back from her angular face and still recovering from the operation, she looked thin and frail. She decided to watch the rescue efforts from a friend's car that was

Inventing TV News

parked nearby. Inside the house, the phone rang constantly. Jeanette answered as many calls as she could. Dave remained as close to the pipe opening as possible. Drillers fired up their clamshell rig and tore into the earth. Everyone knew it was a race to save a child.

***

Print reporters, radio correspondents and cameramen for the papers and the motion picture newsreels were making their way to the scene, eager to match the Times' story. It didn't take them long to find the brown sedan where Alice Fiscus was waiting. Neighbors and relatives near the car were trying to shield Kathy's mom from the gentlemen of the press who were demanding interviews with family members and snapshots of the little girl. Alice changed cars. Johnston saw a man kick a photographer in the stomach. The local beat reporter made his way through the pack and introduced himself to Alice.

"Mrs. Fiscus asked me to help the family," Johnston would later write. "I explained the situation to her and told her there was nothing personal about (photographers') efforts to get pictures; just that they had a job to do and that the story had become much bigger than any one of us."

Alice would later remember only the politeness of the reporters who spoke to her and would downplay any personal knowledge of rude conduct by the press corps. She showed Johnston a snapshot she was clutching. It was a photo of Barbara and Kathy, standing in front of the house, smiling. Johnston circulated the picture among the newsmen and passed along the information tidbits he was getting from

Kathy's mom so they wouldn't have to bother her over and over again to get some human interest for their sidebar stories in tomorrow's papers. The New York Times made space for the story on the front page.

Some of the still photographers were taking pictures of Kathy's sister, Barbara, who watched the digging from a two-seat swing set in the yard. Jeeper, the family's black-and-white dog, kept a vigil at her feet. Alice wished there were some way to get her older daughter away from the press. She gratefully consented when a family friend, Don Metz, an engineer who had come to help with the rescue, invited Barbara to spend the night with his family at their home in Temple City. They had a daughter about a year older than Barbara and they were friends. Jeanette Lyon, Kathy's aunt, decided it would be best for her husband to take their two boys home to Chula Vista.

About 100 onlookers had gathered behind the fences at the edges of the field. At 6:30 p.m. the excavation stopped for a moment. A photographer's camera was lowered into the well in an attempt to document Kathy's position, but the lens fogged. A radio microphone was dropped into the shaft. Everyone listened. There was only silence from the well.

As darkness fell, Kathy's doctor reassured her parents that the little girl must be unconscious. Digging continued into the night. A portable generator powered a battery of klieg lights donated by the movie studios, illuminating a scene as dramatic as a Hollywood film.

***

Inventing TV News

Klaus Landsberg's dinner was getting cold.[6] It was rare for Klaus to be at home for a meal with his wife and son, but he was too busy to eat. The 33-year-old general manager of television station KTLA was on the phone, getting an update on the trapped child in San Marino. Slightly built, with brown hair and eyes and a disarming smile, the self-taught electronic genius was regarded by KTLA staff members with both affection and awe. They gave him nicknames like "Mr. Television" and "The Dictator."[7]

The next morning Landsberg skipped breakfast and worked the phones again. He reacted to the Kathy Fiscus story as both a broadcaster and a father. His son, Cleve, was almost the same age as the little girl in the well. He knew that any parent would feel the same way. To Klaus, Los Angeles was the world's biggest hometown and this was a story the hometown audience would care about. Could he find a way to show the rescue on TV? He prepared to begin another of his typical 14-hour workdays, not knowing that this one would be much longer.

His wife, Evie, watched him leave the house. Klaus worked too hard, she thought. He had never taken a vacation. She also knew it would be useless to protest or even say goodbye as he walked out the door. His thoughts were totally focused on televising the story of the little girl in the well. Evie knew that, to Klaus, nothing was more relaxing than doing the thing that he loved.[8]

Landsberg was born in Berlin in 1916. His parents took pride in their Jewish heritage but described themselves as agnostic. Young Klaus became interested in broadcasting as a child; he was building radios in matchboxes at the age of 9. While still a teenager, he invented a new type of short-wave receiver. At age 20, he helped televise Hitler's 1936 Olympic Games, studied at the University of Berlin and received a degree from the University of Prague. He began working with radar and sonar technology that would enable aircraft to land safely in bad weather; the Nazis classified his work as a military secret, one that had the potential to be a gamechanger in the war. Klaus's family encouraged him to leave Germany and take with his designs with him.

Landsberg arrived in the United States in time to help stage NBC's first public demonstration of television at the 1939 World's Fair in New York. After working briefly for another TV pioneer, Allen B. DuMont, Landsberg was hired by Paramount studios to start a television station on the West Coast. He arrived in Los Angeles by train, with the parts for two television cameras in his luggage.

In 1942 he set up an experimental TV station, W6XYZ, in an old gallery for still photography on the Paramount Studios lot in Hollywood. Curious movie stars stopped by to see what was going on—and find out if this new thing called television really worked. An actor named Dick Lane hosted a variety show called "Hits and Bits." Announcer Jack Latham read the war news for a sparse audience, mostly technical types from the film and aircraft industries who were savvy enough to build their own receivers by hand from spare parts. After the war, the station moved to larger quarters in a converted garage just off the Paramount lot. The garage provided parking space for a truck Landsberg had rigged up to function as a mobile

Inventing TV News

television transmitting unit. The pioneering perfectionist also had the telephone number of just about everyone in Southern California who owned a TV set. Station employees called viewers regularly to find out how the picture looked.

W6XYZ signed on as KTLA, the first commercial TV station west of the Mississippi River, on January 22, 1947. Plenty of Paramount stars were on hand, including legendary director Cecil B. DeMille. But only a handful of "lookers" saw and heard Bob Hope flub the call letters, announcing the new station as "KTL."[9] The first manufactured TV sets were just beginning to show up in appliance stores. At $400 for a 10-inch black-and-white screen, sales were slow.

Virtually all early television programs were live for a few hours per day and originated from a studio or a planned event such as the Tournament of Roses Parade. The bulky equipment didn't travel easily. Early coverage from the scene of a 1943 Los Angeles County Sheriff's Department rodeo involved two weeks of planning and more than three days of set-up. But Landsberg, as one TV writer noted, "believed in television as a window on the world and began bringing the world in, live, to living rooms."[10]

On February 20, 1947, KTLA covered its first news story in the field. It was the aftermath of a chemical blast that ripped through a four-block area in a manufacturing district on Pico Boulevard. The explosion at the O'Connor Electro-Plating Plant later was found to have been caused by a former dairy-farm worker, posing as a chemist with phony credentials. It was the deadliest industrial accident in the history of Los Angeles, leaving 17 dead and 150 injured. Landsberg and his crew lugged their KTLA cameras to the scene and established a signal. Actor-turned-announcer Dick Lane narrated from the edge of a crater 22 feet wide and six feet deep. The blast damaged or destroyed 116 buildings and shattered windows a mile away. A Roman Catholic priest from a nearby church anointed victims with holy oil as Mayor Fletcher Bowron compared the disaster to the 1933 Long Beach earthquake. It was riveting television, but with only about 350 TV sets in Los Angeles at the time, it didn't have much of an impact beyond its tiny audience.[11] Newspapers and radio owned the story as far as the vast public was concerned. If they wanted to see more than a newspaper photograph of the destruction they would have to go to the scene or wait for the film footage to show up in the newsreels at the movie theaters.

But KTLA's live coverage of the factory explosion had proven that television could provide live reports of breaking news, as long as the scene of the story was in a direct line of sight to the station's transmitter. It was located atop a 6,000-foot peak on Mount Wilson in the San Gabriel Mountains, along the northern rim of the valley about 18 miles from Los Angeles. The engineers who took turns manning the facility lived a solitary life on the mountaintop in a two-story cabin. It blended perfectly with the rustic landscape—except for the giant metal tower in the backyard. Landsberg and his chief engineer, Ray Moore, had built the transmitter by hand and constantly tinkered with it to improve the picture. By 1949 the mountaintop crew was responsible for the most powerful TV transmitter in the world, relaying a signal that could reach homes from Santa Barbara in the north to San Diego in the south. It could provide a clear, black-and-white picture to most areas of the Los Angeles basin.[12]

Landsberg sensed he was in the right place at the right time to make history with the Kathy Fiscus story. The scene of the accident was only a few miles due south of Mount Wilson. With any luck, he'd have a straight shot from the little girl in the hole to the transmitter on top of the peak, where the microwave signal could be relayed to Southland viewers. And, by April of 1949, there were more than 20,000 homes and businesses with television receivers in the Los Angeles area.[13]

Landsberg rounded up his staff and began loading KTLA's panel trucks with transmission equipment and two Image Orthicon cameras, each about the size and weight of a large suitcase packed with bricks. Since there would be no prepared scripts, Klaus also needed two announcers who could wing it. He immediately thought of veteran sports announcer Bill Welsh, who had joined the station in 1946. He was described by one TV writer as "probably the fastest man in the business on his mental feet... (never) known to be stuck for an ad lib." Landsberg also put in a call to Stan Chambers, a Navy veteran and USC graduate who had joined the station fresh out of college in December of 1947. An eager young man with a boyish grin, Chambers had filled in on a few news programs and excelled at an improvised man-on-the-street interview show called "Meet Me in Hollywood." Both Welsh and Chambers were reliable members of Landsberg's troupe of versatile on-air performers.[14] Neither was considered an experienced newsman, but they were about to cover a breaking news story that would change their lives.

Terry Anzur

***

The phone was ringing in the ballroom of the Biltmore Hotel. Stan Chambers winced. The annoying bell was interrupting a song by a deep-voiced baritone, the entertainment for about 1,000 ladies at a B'nai B'rith luncheon in downtown Los Angeles. Stan was invited to be the emcee and, like all KTLA personalities, he accepted the unpaid weekend duty as part of the job, publicizing the station's public service to the community. The ringing would not stop. Finally, a waiter picked up the phone and walked in the direction of the head table. How embarrassing to be the person who has to answer that call, Stan thought.

"It's for you," said the waiter.

"Hello?" Stan blushed.

It was his mother. She had taken the call from Klaus Landsberg and had been phoning all over town to find Stan. He could hear the concern, mixed with excitement, in her voice. "Have you heard about the little girl who fell in the well? The station is sending a remote crew there to televise it. You are to meet them out there."

Stan explained to the chairwoman of the luncheon that he would have to leave.

"How will you get out there?" she asked. Stan had taken the bus to the luncheon. He shrugged and said he'd have to make some phone calls. The chairwoman nudged her husband. "Can you drive him?"

As the two men hurried out of the room, the chairwoman took over the microphone. "Stan Chambers has just been called by the station to cover an emergency news story. Let's wish him well and thank him for being with us today." Thunderous applause followed Stan out the door. The chairwoman's husband drove northeast toward San Marino, navigating the streets of pre-freeway Los Angeles County.

As they neared the corner of Santa Anita and Robles, Stan gasped at the scene of what was already a worldwide headline story. He spotted two KTLA trucks in the open field along with a crowd of newsreel and still photographers. He slammed the car door, mumbled thanks, and raced down the dusty, tire-tracked trail that led to a cluster of emergency vehicles and drilling equipment. A circle of men gathered at a hole 30 to 40 feet wide and more than 50 feet deep. At the center, one worker was trying to cut into a metal pipe. Over the din, Stan heard the booming voice of announcer Bill Welsh, beckoning him to the KTLA transmission truck.

"Just in time, Stan. We'll be going on in about ten minutes."

"Is she alive?" Stan asked.

"Yes," Bill replied. "Her mother could hear her crying in the well right after she fell in."

Landsberg stepped out of the truck. Right behind him was cameraman Jimmy Cassin, carrying earphones and a microphone. He gave them to Stan. Bill was already wired. But his forecast of ten-minutes-to-air was premature.

"You made good time, Stan," said Klaus. "This might be a long one."

A long one? Stan considered Klaus a genius but sometimes it was hard to tell what he meant. The TV cameras would take hours to warm up and a long telecast was a gamble.

KTLA aired live programming for only three or four hours per night, never more than 36 hours per week. The cameras were fired up each evening for a few hours and then shut off. Would the tubes give out if they were operated longer?

The topography turned out to be as favorable as Landsberg had anticipated, a direct line of sight to Mount Wilson. But there was no phone available to communicate with the transmitter engineers. Supervisor John Silva called the phone company to order an emergency installation. He was worried when he saw the trucks pull up from KTTV, another commercial station that had recently signed on the air. They would have to race to beat the competition.[15] Bill and Stan walked back to the excavation.

"They started working last night, digging the hole," Welsh explained. "It got deeper and deeper, the sides began to slide and dirt kept pouring down on the guys working at the bottom. There is no shoring or anything to protect the men down there." He pointed to the well casing in the center of the hole. "Little Kathy, she's not even four years old and she's stuck somewhere inside of that."

For a moment, Stan thought he was going to throw up. He had plenty of time to regain his composure because the television engineers were coping with a power failure. The gasoline-powered generator for the cameras and transmission equipment was notoriously unreliable. Stan hoped the problems would be fixed soon. The rescue crews appeared to be getting close.[16]

***

Inventing TV News

At the bottom of the hole and leaning into the lateral tunnel to the well casing, 85 feet down, O.A. Kelly cut into the rusty pipe. The slender, unemployed machinist had worn out a dozen saw blades trying to get through the stubborn metal. Peering into the opening, he could make out something in the darkness.

"I can see something that looks like a dress, but it's too far down to be certain," he yelled to the workers on top. "That's about it. I can't see the little girl."

Loose dirt rained down on Kelly's head. He wished the people up there would stop moving around so much. It was pouring down in steady streams now.

"Let's get him out of there!" someone yelled. The men scrambled, knowing that if they didn't move fast enough they would have a second rescue on their hands. Kelly could be buried alive by a major cave-in.

The dirt held as they pulled him up. But it was clear that this unstable hole was too dangerous. They needed a new rescue plan. It was noon, Saturday, and little Kathy had been trapped for more than 20 hours. "She is everyone's child," Stan thought.

The rescuers who had worked all night were hungry, dirty and tired. And they were starting over again.

The KTLA men regrouped around a second hole being drilled on the opposite side of the well shaft. This one would be lined with 30-inch pipe casing to protect the rescuers from dirt slippage. The TV crew had a continuous wide shot from a camera on top of a truck, while a second camera on the ground had a long cable that enabled the crew to roam the site for interviews and close-ups. Having two different shots also gave the director a backup if one camera failed. But there was still no phone line to reach the transmitter. John Silva scribbled a note and held it in front of a camera: "If you see this picture put us on the air." Minutes later, the on-air monitor in the truck showed that the men on the mountain had gotten the scrawled message. Cameraman Eddie Reznick waved the announcers over to the truck. KTLA would be on the air first!

Klaus Landsberg had one foot on the running board. "We're all ready!" he announced as he jumped into the truck and slid into the director's seat.

The monitor showed the determined faces of men pushing themselves and their machines in a monumental effort to reach the child. It occurred to Welsh that this was nothing like a wrestling match or a baseball game or anything else he had covered before. "What should we say?" he asked his boss.

"Pretend it's a sporting event and give them the play-by-play," Landsberg advised. "I'll have these monitors in front of me to show what is going on and I'll tell you over the earphones what the camera is showing and you just describe it."

Landsberg turned to Stan and said, "Just watch Bill."[17]

At 5:30 p.m. Saturday, more than 24 hours after Kathy tumbled into the well, KTLA finally was on the air with the story. Landsberg made the decision to pre-empt regular programming, including all commercials.

"I'll try to line up some interviews while you go on," Stan told Bill.

Standoffish at first, the workers soon accepted the cameras and the men with the KTLA microphones as part of the team effort. One of the engineers explained to Stan, "We're going to hammer that casing about a hundred feet into the ground to a point below where Kathy is trapped. We have to get out all of the dirt inside the casing, then go down to the bottom and cut a vertical tunnel across, shore it up with timbers and try to dig across to the well pipe where we can get her."

The corkscrew bit from a huge drilling machine was burrowing into the ground, grinding out space for the metal casing to be pounded into the hole. The biggest problem with the plan was that it would take hours to reach the little girl this way.

The announcers worked like a tag-team, taking turns at the microphone. Stan roamed the site for interviews as Bill opened the broadcast, then Stan took a turn on the air while Bill gathered information. Both men adopted an optimistic tone. Over and over, Stan would remind viewers, "Kathy's mother heard her crying in the well, right after she fell in. Her rescuers believe that she is unconscious, oblivious to what is going on."

Under Landsberg's personal direction, the cameras focused repeatedly on the fireman turning the hand crank of the air pump. Stan was often amazed by his boss's eye for significant details. The pump was a powerful symbol of how people were keeping up their hopes that Kathy was still breathing. After a while, Stan stopped identifying the shot. He remained silent and let the picture speak for itself. Would Kathy have to spend another night in the well with no food or water? It was getting dark again.

Although the situation was dramatic, the pictures were monotonous. Forty years later, Stan would remember the black and white images of the "sandhogs" and their machines:

> "At the center of (the) ten-inch screen is a derrick-like piece of heavy equipment illuminated by banks of bright lights. It pounds away on the top of a large cylinder casing, trying to drive it deeper into the ground.
>
> The progress is painstakingly slow. Although the earth shakes as the pile driver hammers away, the huge casing barely budges. One of the volunteer sandhogs steps up to the top of the casing, balances himself on a bucket hanging by a cable and then is slowly lowered down to the bottom of the shaft. He discovers that the casing has hit an underground rock formation, and with his small pick begins chipping away at the boulders."[18]

***

"Every time they try something new, something happens," Alice Fiscus thought wearily. She was dismayed by the setbacks. Never give up, she reminded herself. She was aware of the cameras and microphones, but she saw no reason to turn on the radio and the family did not own a television. She kept her eyes on the activity around the hole. The exhausted diggers sat on cots, blankets over their bare shoulders, each waiting for another turn in the pit, another chance to reach Kathy. The lone bucket was raised, emptied, lowered again,

Inventing TV News

rising and falling like the emotions of the onlookers, all hoping and praying for the trapped little girl.

***

A few miles away at the Metz home in Temple City, the television was on. Nine-year-old Barbara Fiscus was watching. The announcers were talking about her baby sister. The Metz's living room was crowded with friends, neighbors and people they barely knew.

Across town in Sherman Oaks, Myrtle Chizum and her husband started watching the KTLA broadcast with friends who had come over for dinner. Neighbors knocked on the door and asked if they could watch. Soon there was an overflow crowd of friends and strangers in the living room. Hours ticked by and no one wanted to leave the TV set without knowing Kathy's fate. Restaurants and bars with televisions stayed open all night and those without TV closed early. One viewer, Mrs. W.C. Young, called the KTLA switchboard to say that she was counting the buckets of dirt coming out of the rescue shaft and was already up to 191.

Actor Scott Brady, who played tough guys in popular *film noir* movies, was nightclubbing on the Sunset Strip with actress Judy Clark. They turned on the car radio and heard a report about the child in the well. They drove to San Marino and joined the growing crowd at the chain link fence. Stan Chambers spotted the Hollywood faces among the onlookers and interviewed the young stars on TV.

Chambers and Bill Welsh were still taking turns at the microphone. They described the chilly weather and said they had come to the scene without their coats, not knowing they would be here all night. Minutes later, strangers were driving to the site with overcoats and jackets and sweaters. Another man trucked in 70 gallons of hot coffee and stacks of fresh donuts for the rescue crews. Housewives brought homemade cookies, cakes and pies. Klaus Landsberg's wife sent him a change of clothes.

The Harp family on Longden Avenue in Temple City had the only TV on the block, a 10-inch RCA that cost $450. Bread truck driver Clyde Harp, 25, figured that it would entertain his five kids. They were all watching the KTLA broadcast along with their mom, their grandparents, their Aunt Jo and a rotating group of neighbors. Clyde had worked as a cesspool digger to earn pocket money when he was a teenager. He shook his head at the frustrating setbacks in the race to reach Kathy. "I realized the way they were going about it was wrong," he would later say. "They were digging with big machinery and they needed to do it by hand."[19] He made up his mind to join the digging crew and slipped out of the house without telling anyone. He didn't want his family to worry and they were too busy watching TV to notice him leave. Clyde wasn't alone. Before it was all over, more than one thousand men would volunteer for the backbreaking work.

Newsman Cecil Smith had joined the group of reporters, covering the rescue attempts for the Los Angeles Times. He needed a change of clothes and some rest. It was nearly 3 a.m. Sunday as he drove home along Wilshire Boulevard and saw something he never expected. "In front of appliance and music stores," he wrote, "I saw crowds of people, standing in the chill, damp, predawn night, staring at (TV) sets behind plate-glass windows that were carrying the Fiscus story." Smith, who would go on to become the Times' TV columnist, later recalled, "That was the first time I became aware of the potential of television."[20]

One of the faces pressed to the glass that night belonged to young Joel Tator. He had been to a movie theater on Hollywood Boulevard with his family and they were walking home. Joel lingered in front of an appliance store, transfixed by what was happening to the little girl in San Marino. It was a sight that would change the trajectory of his life. Joel would grow up to become one of the most innovative directors of live television programs in Los Angeles, many of them on KTLA.

George Putnam's life also changed that night. Like many radio announcers, he looked down on the upstart medium of television. Some refused to accept TV assignments. Putnam was a network radio star in New York and could also be heard on Fox Movietone newsreels. According to Putnam, the Fiscus telecast, "convinced many radio newscasters, myself included, that this new electronic medium was the way of the future."[21] He moved to Los Angeles in 1951 and became a news anchor on KTTV.

Nationwide network television was only an idea in April 1949. No cables existed to send the television signal to other parts of the country. The rest of the world was listening to the radio or reading the press accounts for the latest on Kathy's fate. Newspapers in Sweden, England, Australia and across the United States held the presses for the latest on the trapped child. The rescue effort made front-page news in the New York

Times for three days. Callers jammed the newspaper switchboards in Salt Lake City and Chicago. One man poured out his heart to the operator at the Pittsburgh Post-Gazette: "I'm the father of three little ones and this story about poor Kathy really hit me." Only in Southern California could people share these emotions in front of television sets, watching as men who dug cesspools and sewers for a living became folk heroes in the race to save a helpless child.

"You talked to the workers coming up," Stan remembered later in an interview. "You described the digging. People came in and volunteered. They were lowered in buckets and the whole evening all you saw was a bucket come up and dump its load. These people would go down and dig for half an hour or 45 minutes—perhaps a dozen of them—and that was all you saw, but every time someone would come up, he was completely exhausted, sweating, just overwhelmed by the situation."[22]

The KTLA crew had lowered a microphone into the shaft. Everyone watched and listened. First, a scratch, the pick cutting into the rock. Then, the scrape of the bucket against the side of the reinforcing pipe. And then...singing?

***

"Big Bill" Yancey sang as he scraped away the sand and chipped at the rocks. The 38-year-old cesspool contractor didn't care that he was loud and off key. It made him not think about how hot it was in the hole, how cramped he felt. He didn't know that the TV announcers were speculating about the very same thing. How long could a man stay down there and how could a little girl survive?

Yancey had been an underwater demolition expert for the Navy during the war, clearing enemy beaches for American troops. After more than two hours underground, he had managed to single-handedly clear five feet of tunnel and asked to be hauled to the surface. Excitement rippled through the crowd, then cheering broke out at the sight of Yancey emerging from the hole, covered with sweat and dirt. He staggered to the first aid area and collapsed onto a cot, all within view of the cameras. When he was feeling better, he told Stan Chambers—and everyone watching on television— what it was like in the rescue shaft, where a dangling thermometer measured the temperature at 90 degrees.

"Hot down there, really hot," Yancey said, munching on a sandwich. "The big trouble was the rocks, big as your head, some of them, and hard to handle in that space. I could have stayed down there longer, but I was tired and I figured that a fresh man could do the work faster."[23]

As Sunday morning dawned, the buckets were bringing up wet sand. The crowd cheered for Herb Herpel, the next hero to enter the pit. The ex-Navy Seabee was Big Bill's business partner and the father of a 10-month-old boy. His wife, Nancy, proudly looked on as he balanced himself on the tilting bucket and grabbed hold of the cable. When he reached the bottom of the shaft he did not sing.

"Hell! Damn!"

Now KTLA's microphone was picking up swear words. And suddenly, a scream.

"I've hit water," Herpel yelled to the surface. "It's coming in on all sides. Pull me up!" Men scurried around the rescue shaft, hoisting Herb above the flooded area. "That's far enough," he shouted over the microphone. "Let me check this water level... it's just oozing in... plenty of mud... but we'll just have to haul it out."

Television viewers saw and heard every heart-pounding moment of this close call. The crowd at the scene cheered Herb's return to the surface. He would be all right. But experienced excavators knew this was the worst news yet. The rescue shaft was now 100 feet deep and the water seeped in just as the men were digging the lateral tunnel to Kathy. Was the child under water? Pumping equipment, 120 feet from the hole, began churning at full capacity. The water eventually filled up a nearby reservoir and had to be diverted to a second catch basin.

The sandhogs worked in two-man teams; one man digging while another filled the bucket. Clyde Harp took his turn in the shaft, swearing like a trooper but not because of the water, sand and mud. He had just sent up a bucket of tools and the guy on the surface had dumped the tools back into the shaft, narrowly missing Clyde's head.

"Quit swearing down there," yelled Mark Nottingham, Clyde's friend and former boss. "You're on TV!"

Clyde worked in the pit as long as he could and had to be helped to a stretcher when he finally returned to the surface. Bill Welsh was there with a microphone to ask why he had volunteered. "Over in Temple City where I live," Clyde said, "I have a girl six weeks old, a girl two years old, a boy three, a girl, five, and a boy seven."

A loud cheer went up from the family in front of the TV set at Harp's house on Longden Street. It was their hero—Daddy![24]

***

People in the crowd whispered that the workers could hear Kathy crying. But when it was Whitey Blickensderfer's turn to go below, all he could hear was the loud dripping of water, like bullets hitting the iron casing. He had to be pulled up right away. The water was rising too fast and the mud was up to his thighs. The men held a conference by the hole. They knew they could not afford to give up now. Blickensderfer returned to the pit. The 43-year-old ex-hard-rock miner, ex-boilermaker, and ex-sandhog was unemployed and had driven across the Southland from his home in Reseda to volunteer. He didn't tell anyone about the pain from his hernia. Whitey and his wife could not scrape together enough money for the operation to get it repaired.

Between interviews with the sandhogs, Stan Chambers kept warm in the cab of a pick-up truck. He was still narrating the pictures Klaus Landsberg described over the earphones. Below the surface, workmen were digging a lateral tunnel from the rescue shaft to the pipe that held little Kathy, only about five feet away.

Suddenly, the roof of the tunnel collapsed. There was another flurry of activity as two workers were pulled out from under three feet of soggy sand and given emergency treatment.

The tunnel would have to be shored up with lumber and that meant another delay. "This is ridiculous," Stan thought. He still wondered if anyone was watching. To his amazement, the tube cameras that were only supposed to last a few hours were still transmitting television pictures after being on all night.

***

Palm Sunday churchgoers prayed for Kathy. The crowd at the fence swelled to more than five thousand as local residents decided to visit the rescue scene on the way home from morning worship services. Years later, Kathy's mother would credit the TV coverage for keeping the crowd from growing even larger. Thousands more were praying at home or wherever they could find a television set.

Marie McLoughlin was watching the KTLA telecast with her mother at their ranch in the Ventura County community of Oxnard. She made a fresh pot of coffee and poured two cups.

"Marie, look at that nice fellow on television," her mother said. "He is so concerned, so emotionally involved with the rescue. He must be a nice person."

Marie recognized Stan Chambers. She had attended the University of Southern California with him. Although Marie was newly married to her college sweetheart, she wondered if Stan was still single. He would be the perfect match for her sister, Beverly.

***

Hours passed by as the workmen inched closer to the child in the well. The bucket traveled up and down, up and down, bringing out more mud and rocks and serving as an elevator for the sandhogs. Sunshine beat down on the field, raising the temperature in the shaft. The hot, sticky weather didn't wilt the enthusiasm of the crowd, still cheering for each sweaty miner to emerge from below. They were still digging by hand because electrical tools posed a deadly hazard in the water-filled hole. If she was still alive, could the little girl hold out much longer?

Clyde Harp knew of a construction supply house in South El Monte. Like most businesses it was closed on Sunday, but he tracked down the owner and persuaded him to loan his pneumatic grinder to the rescue effort. Sirens wailed as Clyde returned to the scene of the accident with a police escort. By late afternoon, the KTLA microphone picked up the screeching of a drill as Clyde began cutting into the pipe. It was obvious to Stan that workers were getting very close.

All of sudden, no one wanted to be interviewed.

"Officials are grim and some are testy," Stan would later write. "For the first time, the microphone in the casing is turned off. It is our last official contact with the men below. No reason is given, but it quickens fears that things have turned for the worse."

Stan's heart went out to the little girl's parents. He wondered how they were holding up under the pressure but made no attempt to walk over to the house and invade their privacy. Kathy had now spent 48 hours underground, alone in the dark.

Terry Anzur

***

At 6:03 p.m. Sunday, O.A. Kelly and Whitey Blickensderfer reached Kathy through a 12-inch by 22-inch hole in the rusty iron well casing. They could see the child tightly wedged just below them, in mud. They relayed the news to Raymond Hill, the engineer in charge. But Hill refused to say whether the child was dead or alive. The workers called for ropes, hooks and a can of grease, anything they could use to free the child from the pipe. Kathy's parents came out of the house and were escorted to a waiting police car. Dave sat in front with the driver and Alice shared the back seat with her sister, Jeanette. Officers cleared the streets leading to St. Luke's Hospital. Motorcycle cops revved their engines, ready to provide an escort.

A white canvas bag was lowered into the hole and raised again, empty. The crowd fell silent. For the first time since Friday evening, firefighters stopped cranking the air pump. Blickensderfer emerged from the shaft and directed the operations on the surface for as long as he could. Exhausted and in pain from his hernia, he was taken to Huntington Hospital. Next out was Kelly, who stumbled to a first-aid cot. He had touched the girl but was unable to free her legs. He still could not tell if the child was dead, or simply unconscious. There was another tense conference next to the hole. At 7:20 p.m. the firefighters resumed their duties at the air pump. It was getting dark again.

One hour later, Yancey entered the pit with Kathy's pediatrician. Dr. McCullock, wearing dungarees and an aviator's cap, was lowered into the rescue pit with a parachute harness. Hopes flickered and faded minutes later as the grim-faced doctor was pulled up again. Klaus Landsberg told Stan Chambers over the earphones to get ready for an announcement. The KTLA cameras maneuvered into position. Bill Welsh handed Stan the microphone and stepped away, conferring with Los Angeles County Sheriff Eugene Biscailuz. It was 8:45 p.m. as Stan watched Bill walk toward the Fiscus home. Kathy's parents had left the police car and were waiting inside the house.

Alice Fiscus remembers hearing the dreadful news, not from the TV announcer, but from the family doctor. A minister and others offered condolences, but it was all a blur for the grieving mother who reportedly was under sedation. Years later, she would not remember speaking with Bill Welsh. But the TV announcer would never forget his visit with the rest of the family. He was told that some of the relatives had seen him on TV.

"I told them I was sorry to be the one to tell them that little Kathy was not coming home again," he would later tell interviewers. "To tell a whole room full of aunts, uncles, cousins and know that what you have to tell them will emotionally impact them is worse than reporting the news (on TV) because you saw your audience."[25]

The waiting ambulance would be needed after all. Kathy's uncle, Ham Lyon, collapsed and was rushed to St. Luke's Hospital.

Now that the family had been notified, it was time to tell the world.

Terry Anzur

***

At 8:58 p.m. Dr. Paul Hanson, the family physician who had delivered baby Kathy, picked up a public address microphone. He stood in front of the TV cameras.

"Ladies and gentlemen," he said, "Kathy is dead and has apparently been dead since she was last heard speaking Friday." The doctor explained that the apparent cause of death was drowning or suffocation. He continued, "The family wishes to thank one and all for your heroic efforts to try to save our child..." He asked the crowd at the fence to show their respect for the family by leaving the area. "If this had been your child, we are sure you would not want a crowd remaining at the scene of the tragedy."

Slowly and sadly, the onlookers moved away. The Associated Press flashed a one-line bulletin: "They found Kathy Fiscus dead."[26]

For Stan Chambers, the story was written on the faces of the tired and dirty men who had toiled for more than 50 hours. He would recall the sight of tough guys openly weeping and consoling each other: "The brave men, who worked desperately to reach her, broke down and cried over the death of the little girl they never knew but now would never forget."

Bill Welsh took his last turn at the microphone. He spoke of the "unpleasant duty" of notifying the family but refused to reveal further details because it would be in "very poor taste." It was up to Chambers to bring the television coverage to a close, more than 27 hours after it began:

"And now it is 9 o'clock Sunday night, probably the longest television broadcast in history. And we're sorry that this is the way we have to sign off, because we always hoped that we would have had a happy ending. We want to thank you for staying with us during these long, long hours, and for being with us. I know the family feels the same way, and appreciates the sorrow as you've expressed. And so, ladies and gentlemen, we leave San Marino... hoping to have given you the service that we wanted to. And now we return you to our studio."[27]

***

Welsh saw no reason to stay on the air any longer. "Television was a little more sensitive to people's feelings in those days," he would later recall. "We did not stay on the air to watch them bring up the body."[28]

The first man to enter the rescue shaft would be the last man out. The telecast had been over for nearly an hour when "Big Bill" Yancey made the final descent. He came up cradling Kathy's tiny body in his muscular arms. The frilly edge of her pink party dress could be seen under the gray blanket covering her body. A newspaper reporter noticed that the Kathy's dog, Jeeper, was still keeping watch in the yard by the swing set, his head resting on his paws as if looking for his little playmate to come bouncing up the path. Kathy left the scene of the accident in a hearse that transported her body to a nearby funeral home.

Terry Anzur

Mrs. Burdette Cogswell didn't realize how tired she was until she walked home. A member of the San Marino Canteen Corps, she had worked at the Red Cross table for 38 hours, serving sandwiches, drinks and apple pie to the volunteers. Many of the workers had not slept in two days, but they did not leave. Raymond Hill led the effort to clean up the site, vowing that by dawn on Monday "no one will know there has been any activity here but plowing."

***

Doctors spent the next several days responding to rumors about Kathy's death. They flatly denied reports that a rescue rope had strangled her. She was found sitting up in the pit, legs curled underneath her, arms at her sides and a rope coiled around her waist. A coroner's autopsy would reveal no broken bones, no cuts that could have caused a loss of blood. The doctor who performed the post-mortem said that no water was found in Kathy's lungs. The heat and dampness in the pipe may have accounted for her soggy dress and the wrinkled condition of her hands and feet. She had suffocated, doctors said, and apparently suffered little pain.

"It was the saddest day of my life," Clyde Harp would say of the failed rescue effort. "And the second saddest was the day of the funeral." Wearing their best Sunday suits, the sandhogs posed for a group photo three days later at the memorial service for the child they had tried so hard to save.[29]

Bill Johnston, the Los Angeles Times newsman who broke the story, received a "reporter of the month" bonus: ten dollars. Recognition for the would-be rescuers came from the local

Inventing TV News

television audience. A doctor who had been watching the telecast volunteered to operate on Whitey Blickensderfer's hernia, free of charge. Upon hearing that some of the brave men were unemployed, others offered jobs. The mayor of San Marino and the Chamber of Commerce set up a fund for public contributions to reward the group of men who "without thought of compensation and with utter disregard of their own comfort and safety, applied their talents and strength to a monumental battle with death." Kathy's family received numerous donations and turned over all the money to the rescue fund. One worker used his share of the money to buy a TV set for little girls in a tuberculosis sanitarium in nearby Duarte. The children named their television "Kathy."[30]

***

Klaus Landsberg realized what KTLA had accomplished. He returned home, tired and unshaven, and told his wife, "This is tragic, but this is also television history."[31]

It was the first time the new medium had scooped newspapers, newsreels and radio by showing the story as it happened. Landsberg got the credit for the technical breakthrough and KTLA became the television news station of record in Los Angeles. As a result of the marathon broadcast, the engineers learned that continuous use was better for the cameras and transmission equipment than the wear and tear of turning them on and off each night. An editorial in Daily Variety praised the KTLA remote crew for delivering compelling pictures and commentary "on a par with any big dramatic screen or stage presentation."[32]

Many Southland residents did not know, until they read about it in the Los Angeles Times, that there was a second television station on the air during the failed rescue effort. KTTV, jointly owned by the Times and CBS, carried a live signal from three cameras on the scene beginning at 6 p.m. Saturday. By the time announcers Walter Carle and Bob Purcell signed off the air at 9 p.m. Sunday, they had been on the air continuously for 27 hours, almost as long as KTLA. The Times duly reported television's "fire trial," but without even mentioning Klaus Landsberg and his crew, setting up a news rivalry between the two stations that would continue for decades to come.[33] According to Stan Chambers, KTTV had been on the air for only three months and its picture was of a lesser quality, leading viewers to develop "the KTLA habit."

For Klaus Landsberg, live news coverage was a way of differentiating KTLA from other TV outlets in Los Angeles, network TV programming and even motion pictures. He didn't need to worry about ratings. The 1949 Hooper index—the early counterpart of the Nielsen ratings—showed KTLA with nine of the top ten programs in Los Angeles.[34]

But in 1949, manufacturers were still selling the public on the idea that television was a necessary appliance in the home. Early KTLA remotes placed live cameras on busy shopping streets for "Philco Days," as if to reassure buyers there would be something to watch if they made the $400 investment in a receiver. The Mirror's radio-TV column pointed out that "TV dealers were quick to rig up sets in store windows" tuned to KTLA Channel 5. In terms of transforming television from a novelty to a necessity, the Fiscus telecast was a huge success,

Inventing TV News

credited with the sale of 100,000 television sets in a single weekend. The owner of a San Francisco appliance store later told a trade publication that the story of Kathy Fiscus gave consumers "an idea of what TV could really do and many were hooked—from that point the sales blitz was underway."[35] By 1950 there were 300,000 receivers in Los Angeles.

Kathy's socioeconomic status, as the child of a white, upper middle-class suburban family, represented the primary audience for the TV sales pitch, along with the upwardly-mobile blue-collar workers of the post-war era, represented by the rescuers. However, newspaper commentators at the time did not appear to perceive any commercial motives. They assigned a more noble significance to the event. An article by Elaine St. Johns in the Los Angeles Mirror compared the solidarity of the rescue effort to the determination that enabled Americans to win World War II:

> "... here the world united over the life of one child. That, perhaps, is the miracle of little Kathy Fiscus. No color, race, creed, union or non-union, rich or poor distinction arose to mar the united efforts of the men and women who fought to save her.
>
> Headlined in foreign lands was this single story that does much to wipe clean a few of the other headlines... headlines written when men united for destruction. Life is not held cheap in America!"[36]

***

The same theme could be found in other American newspapers, including this passage from an editorial in the New York Times:

> "The world is overladen with great problems. Two great wars and many smaller ones have cost the lives of multitudes, including little girls as dear to their parents as Kathy was. But we still know, even if the mad theorists at the other side of the world do not, that one life—one tiny life—is beyond price."[37]

Beyond the technical and commercial breakthrough, the broadcast also introduced the audience to a new stranger in the living room: the television newsman. Bill Welsh's fame grew to the point where received a "permanent contract" from rival KTTV in 1951. He went on to broadcast a total of 49 Rose Parades and cover 63 sports, ranging from lawn bowling to professional wrestling. He emceed televised beauty contests and Easter sunrise services, preferring live broadcasts to the scripted news programs he considered "boring." But there was never another story to match the impact of the Kathy Fiscus rescue effort.

Welsh later wrote, "The public was so moved by this event that they felt they wanted to share everything with me because I had been their surrogate at the scene of the rescue effort. From that time on, the public realized that television was much more than 'home movies.' It was a thing with a heart and a soul and it was going to have a tremendous impact on their lives."[38]

Stan Chambers became a local celebrity as the result of the Fiscus coverage and spent more than 50 years at KTLA. He accepted Marie McLoughlin's invitation to dinner and married her sister, Beverly, soon after. They had eleven children. Their daughter, Maggie, says the Chambers family always thinks of Kathy Fiscus as "the little angel who brought our mom and dad together."[39]

Chambers lost his voice and was unable to work for several days after the marathon broadcast from San Marino. He had plenty of time to read his fan mail and reflect on the impact of Kathy's story.

One viewer, Stewart Stern, praised the telecast's inspiring portrayal of "people at their best." He wrote, "We had been to a party. We sat on the floor in our tuxedoes and watched you through the night and into the next night. Until last night, the television set was no more a threat to serenity than any other bit of furniture in the living room. Now you have utterly destroyed this safety forever... you and the epic of which you have been a part of this weekend have made us know what television is for..."[40]

***

The child's death gave an immediate push to the bill her father had lobbied for. The Daily Mirror reported on April 11 that "the state legislature today rushed passage of a law that will make another such accident impossible." One month later, California Governor Earl Warren signed a measure to require capping of abandoned wells. This was a great comfort to Kathy's grieving mother. "The main thing I've always said is whenever this is brought to light, this reminds people to be very careful and to report everything they see in the line of a hole or an open ditch," she told a local newspaper on the 50th anniversary of Kathy's death.[41]

But there would be others. And each time another boy or girl was trapped underground, reporters would contact Alice Fiscus, the first mother to endure the deeply personal loss of a child in full view of television cameras. She turned down most of the requests. Although barely aware of the televised coverage at the time, she would later acknowledge that "the impact from TV and radio was overwhelming. We kept hearing about it from all our friends and gradually the world. It was hard to comprehend until all the mail started coming in by the hundreds, from every country except Russia. I think the TV, where people could actually watch, definitely made the grieving easier for us because most everyone that knew us also knew of what and when things were going on. Explanations were not necessary."

In 1987, Kathy's mom again warned parents of the potential danger posed by uncapped wells. Her comments were prompted by the successful rescue of Jessica McClure, a toddler who fell into an abandoned well in Midland, Texas. Alice Fiscus had watched the television coverage of someone else's little girl being lifted to safety and reunited with her parents. "You get to the place where you have a stone wall, where you see the advantages of things like this that take place, rather than the hurt that comes with it," she explained to an interviewer.

Inventing TV News

***

On April 13, 1949, Kathy Fiscus was buried in Chula Vista, her mother's hometown. Fifty years later, a tennis court at San Marino High School has been built over the place where she died. Nearby, under an oak tree at the edge of the baseball field, is a small stone marker. Few visitors venture close enough to see the brass plate. It reads: "...the miracles of radio and television transported our entire nation to this field. Here we witnessed the courageous attempts of so many to save a little girl. This should not be a spot to recall sadness. This should be a place to visit and say prayers."

Across town, children leaving the San Marino Public Library pass by a rose bush with delicate pink flowers, the color of a child's party dress. Under the bush is a plaque for Kathy, "a little girl who brought the world together—for a moment."

Hardly anyone notices the inscription. Perhaps they are hurrying home—to watch television.

Chapter Two: Covering Crime

Television's coverage of the tragic death of Patricia Jean Hull...brought home the realization that a real-life story still packs more punch than any staged show.

—Walter Ames,

Los Angeles Times Columnist
May 26, 1951[42]

In the years following the Kathy Fiscus story, the growing number of Southern California homes with TV sets could expect to receive coverage of natural disasters, politics, celebrities and sports. However, the 1950s audience could scarcely have imagined the degree to which local crime dominated TV newscasts by the 1980s, when stations would be criticized for the unspoken motto: "If it bleeds, it leads."[43] News competition in television's early days was driven by the commercial necessity of selling more TV sets to consumers and selling advertising time to potential sponsors. Stations aired money-losing news programs in order to fulfill the licensing requirement for public service. Sordid details of violent crimes were seen as bad for business and "a subject too morbid to be called a public service."[44] That assumption would be challenged in 1951 by the abduction of a little girl from a suburban Orange County movie theater during a children's matinee.

***

Disaster was part of the program from the earliest days of television broadcasting in Los Angeles. In 1933, experimental station W6XAO televised rapid-process film of the aftermath of the 6.3 magnitude Long Beach earthquake, in which 115 people were killed. Although the city had only a few receivers, viewers could see pictures of the destruction while access to the disaster area was still restricted. Television's potential for timely news reporting was established in Los Angeles years before most cities had even one station on the air.[45]

The technical genius behind W6XAO was Harry Lubcke, a young protégé of electronic television inventor Philo T. Farnsworth. "The eventual goal is a nightly spot newsreel service based on topics of local interest," Lubcke predicted. "This would be telecast in the early evening when the family is at home and at leisure to receive the events of the day."[46] But Lubcke's boss wasn't interested in selling news to the home audience; Cadillac dealer Don Lee, the owner of W6XAO, demonstrated television in his downtown automobile showroom to attract the same upscale consumers who might be interested in purchasing luxury cars.[47]

Early broadcasters also invested resources in covering politics. Philadelphia's experimental W3XE became the first local station to cover a political convention as Republicans nominated Wendell Willkie in 1940. Believed to be the first remote news broadcast, the program was also carried to some 500 sets in New York via coaxial cable.[48] Most of the 150 viewers in Philadelphia were executives or engineers for the manufacturing company that owned the station. But Philco wasn't motivated by journalism or public service. It provided programming and set up displays in appliance stores to encourage customers to buy Philco receivers. TV pioneer Allen DuMont had the same feeling about the 1949 debut of his WDTV in Pittsburgh, calling it "just an installation to sell some sets."[49]

Except for the sales of actual receivers, there was no possibility of commercial profit in the early days of television broadcasting. Investing in a local TV station was a leap of faith that required deep pockets and technical know-how. By the end of World War II in 1946, nine applicants in Los Angeles were jockeying for the seven available television channels. Topping the list were NBC and ABC, broadcasting companies with established radio networks. CBS affiliated with the station licensed to the Los Angeles Times and another permit went to Dorothy Thackery, the publisher of the New York Post who already owned three radio stations. Other successful applicants were Cadillac dealer Don Lee, and Earle C. Anthony, the leading Packard dealer on the West Coast.

Paramount's W6XYZ completed the list of seven brides for nine brothers. In choosing, the Federal Communications Commission (FCC) apparently favored broadcasting experience over ties to the local community. Applications by the Riverside-based Broadcasting Corporation of America and the Hughes Tool Company were denied, with the reclusive

Inventing TV News

mogul Howard Hughes losing out because he reportedly refused to attend the FCC proceedings in person. Although profits remained a distant dream, the priorities of early licensees were in tune with the first electronic TV image broadcast from Farnsworth's San Francisco lab in 1926: a dollar sign.

Paramount executive George Shupert told Television Magazine the film studio was investing in a Los Angeles TV station to pitch products to the Hollywood elite. "The first sets... will be in the hands of the top people in motion pictures—the people who set styles and the pace for the rest of the country. If they can be sold on a product, by use of a visual means and start using it in pictures, impact on merchandising would be terrific. Thus, 1,000 sets in the Los Angeles area may be more profitable than 20,000 or 30,000 elsewhere."[50]

Throughout the 1940s and 50s, most local stations had some form of news. But they didn't try very hard to speed up delivery of newsworthy pictures to the home. The Chicago stations all had remote trucks for sports and special events but rarely used them for local spot news.[51] Pittsburgh's WDTV hired an outside production company to shoot film footage for a nightly, five-minute local newsreel called "Pitt Parade" because the station debuted in 1949 with no studio and no news department of its own.[52]

Even as commercial TV started to become profitable, stations did only enough news to prove they were meeting a government mandate to act in "the public interest, convenience and necessity."[53] Deep-voiced announcers read radio-style headlines over a slide that said "news" or voiced over silent newsreel footage with background music. The pictures might come from around the world but rarely from down the street. Sometimes the "news" was a week old. Established producers of weekly motion picture newsreels doubted that local television crews could find enough suitable material to fill up a daily show. Like the proprietors of movie theaters, early station owners didn't want controversy to interfere with entertainment. They preferred newsreels that focused on the sensational and offbeat while avoiding social and political issues.[54]

"There were pictures of important people arriving or departing, heads of state making speeches or reviewing troops," recalled veteran KTLA reporter Stan Chambers. "There were horse races and beauty contests, hurricanes, flood damage and assorted light features. In those days newscasts were routine and were not expected to give the station big ratings. We presented the news because it was our duty to the viewers and the (FCC). We did not do it to build big audiences."[55]

An unidentified station manager in the 1950s summed up the prevailing attitude this way: "Get through with the news and get back to the wrestling."[56]

KTLA found a way to cover the 1952 elections without interrupting one of its most popular programs: Ina Ray Hutton and her all-woman orchestra. The show was re-titled "Election Jamboree" and expanded to two hours. The stunning female bandleader fumbled through the returns as popular World War II General Dwight Eisenhower easily defeated Adlai Stevenson. Hutton made an exception to her strict rule of barring men from the set while her gorgeous musicians were performing. KTLA newsmen appeared briefly to update the local and statewide election results. The "Ike and Ina" show was a hit with the audience, beating all the networks and local stations in Los Angeles.[57]

Like other Los Angeles stations, KTLA also aired regularly scheduled, studio-based news programs, usually 15 minutes or less including sports and weather. Then it was on to sporting events or programs like "Pantomime Quiz" and "Queen for a Day." Some KTLA pioneers were convinced that Paramount studios dabbled in TV strictly for the tax deductions[58] and intended to produce the bare minimum of news to qualify for a license. The FCC tended to favor the TV network-building agenda of East Coast radio broadcasters like NBC's David Sarnoff and William S. Paley of CBS. Paramount needed public service programming to convince skeptical regulators that Hollywood would not monopolize television to promote its stars and films.[59] But KTLA general manager Klaus Landsberg intended to deliver far more than the bare minimum with Paramount's bankroll. He did not believe news should be limited to a studio-based combination of radio-style scripts and newsreel footage. He was eager to build on the success of the Kathy Fiscus broadcast. The following week KTLA cameras were on the scene of a dam collapse in Santa Monica where two people died.

This was in marked contrast to the trends in other early TV markets. As former CBS News president Sig Mickelson has noted, viewers in the East and Midwest "had no event matching the drama of the Kathy Fiscus story."[60] New York was developing prototypes for network newscasts with Lowell

Thomas, Douglas Edwards and John Cameron Swayze in the anchor chairs. From 1949 to 1955, the "Chicago School," emphasized folksy, unscripted, studio-based productions that launched the career of future NBC "Today" host Dave Garroway. [61] Stanley E. Hubbard's KSTP in Minneapolis boasted the nation's first scheduled 10 p.m. news, seven days a week, but it was the typical blend of radio-style copy with local and national film. Led by KTLA, Los Angeles developed its own 'school' of news coverage, sending its camera crews out of the studio and emphasizing live, on-the-scene local reports.

"Not content with merely standing by for something to happen, the station sends its trucks out looking," one reviewer wrote. [62] KTLA interrupted scheduled programs when necessary to cover everything from sunrise services to train wrecks. It would eventually bring the ultimate destroyer—the atomic bomb—into the living room. But even Klaus Landsberg hesitated when it came to a shocking, violent crime against an innocent child.

***

Parents in the early 1950s didn't rely on television as an electronic babysitter. Children's programs like KTLA's "Time for Beany," or Chicago's "Kukla, Fran and Ollie," and the wildly popular "Hopalong Cassidy" westerns were prominently featured on early TV channels but only for an hour or two around dinnertime. Parents who wanted to keep the kids occupied for an entire Saturday afternoon didn't have to buy a TV. They could send the children to the movies.

Inventing TV News

It was May 19, 1951. The Saturday "kiddies' matinee" at the Valuskis Theater started at 1:30 p.m. Ten-year-old Patty Jean Hull was going to the double feature with her two younger brothers, 7 and 8. They could walk to the movies because the theater was only about a block from their home in Buena Park, an unincorporated town on the western edge of suburban Orange County, about 20 miles southeast of Los Angeles. Patty asked her mom for some extra change to buy double ice cream cones. Terry Hull gave her daughter the money and waved goodbye as her three little ones skipped down the street.

The children bought their tickets and looked for seats in the theater. Patty found one next to a friend from her fourth-grade class at Lindbergh Elementary School, 10-year-old Irma Shaw. A stranger sat down next to them. "He bothered us, so we had to change seats," Irma would later recall. The theater darkened as the movie began. Patty's brothers enjoyed the show but couldn't find their sister when it was over. They headed home, thinking that maybe she had left with one of her girlfriends. They were half right.

Patty had left the theater, but not with a friend.

At 6 p.m. Patty's parents sent the boys back to the theater to look for her. Then they searched the neighborhood but found no trace of their daughter. At 9 p.m. they called the Orange County Sheriff's Department. Patricia Jean Hull was gone.[63]

***

Terry Anzur

Roy Reynolds and his wife, an older couple from Downey, were passing time in their parked car outside the theater at 5:30 p.m. They had come to visit their daughter who lived nearby and she wasn't home. They noticed something strange. A little girl in braided pigtails and a red sweater appeared to be struggling with a stocky, dark-haired man who was dragging her down the street. Was it a father disciplining an unruly child? They couldn't be sure.

At a motor-court motel in Buena Park, 'Holly' Holland was relaxing, looking out the window of the one-room cabin he called home. The 28-year-old house mover was surprised to see the man who rented the neighboring cabin standing outside with a young girl. Holland noticed the child's pigtails and her red sweater, blue jeans and bobby socks. He saw his neighbor fumble with the keys. "He took the girl in and shut the door," Holland later told police. "I didn't see them again."

Several hours later, a used car salesman showed an automobile to a man who didn't give his name. It was closing time and the nervous customer seemed to have a hard time making up his mind about buying the car. He said he'd have to think about it.

It was 9 p.m. when Henry Ford McCracken walked into the White Elephant Café. Lee Stradley, the tavernkeeper, recognized McCracken as a customer who sometimes came in and played the guitar for tips. But McCracken seemed jumpy. He said he didn't have his amplifier and needed to borrow a car to go get it. Stradley didn't really want to loan his car to a man he barely knew, but McCracken swore he would only be gone for ten minutes. The bartender reluctantly handed over his keys. Five hours later—at nearly 2 a.m.—McCracken returned the car. He offered no explanation, but Stradley noticed he was "nervous as shit."

***

By Sunday morning, everyone who knew Patty Jean Hull's family was talking about her mysterious disappearance from the theater. The father of a 10-year-old boy came forward and said a stranger had made "improper advances" to his son during the movie. Two other little girls, 9 and 11, told their mother about a man who had "bothered" them in the same theater a week earlier. The two young sisters gave a description that fit the man who had been seen dragging a child down the street outside the theater.

All the clues led investigators to McCracken's cabin at the motel. Inside, the room was splattered with bloodstains, so fresh they were still wet. Nearby, they found a razor-sharp paring knife with its shiny blade scrubbed clean. A detective spotted a single blonde hair that might have belonged to the missing child. Investigators also found blood-spattered clothing, a man's green sports outfit. It was another match with the descriptions of the person who had annoyed children at the Saturday matinee. Deputies arrested McCracken when he returned to the motel on Sunday night and took him to the Orange County jail in Santa Ana.

Detectives didn't believe the suspect's explanation that he had cut his finger. There was too much blood. They also doubted his story about having left the theater for a nap and a date with a woman who never showed up. He didn't have a plausible alibi for those five hours of driving around in the bartender's car, except that he was looking for an old friend whose name and address he didn't remember.

And then there was his record. The newspapers reported that the 34-year-old McCracken had been arrested seven times, including one charge of suspected draft evasion in 1944. After an honorable discharge from the military in 1946, he was arrested on a charge of child molestation in Santa Ana and disappeared, forfeiting bail. He next turned up in Detroit, suspected in another child molesting case that was dismissed. By 1949, McCracken had been picked up for prowling, contributing to the delinquency of a young girl, and an attempted sex attack. In January 1949, he was convicted of contributing to the delinquency of a minor and sentenced to 90 days in the House of Detention in Plymouth, Michigan. Authorities were drawing up papers to have him committed when McCracken fled the state. He returned home to Santa Ana where he was picked up for annoying little girls at a birthday party. He served six months in an Orange County jail road camp for failing to register as a sex offender. He was released less than two weeks before Patty disappeared.

The suspect's 72-year-old mother was not surprised at his arrest in the Patty Jean Hull case. "Men like my son should not be allowed loose," she told a reporter from the Examiner. "I cannot tell you how sorry I am for the mother of that poor little girl in Buena Park."

Deputies relentlessly interrogated the suspect but McCracken wouldn't talk. They had a known sex offender in custody and a bloodstained crime scene—but no victim. Where was Patty? And was there any chance that she could still be alive?

***

A search headquarters was set up at the Buena Park fire station. It seemed that nearly every family in the town of 5,000 had sent a father or son to help look for the missing child. Mothers and daughters worked overtime in their kitchens and set up a mess hall in the auditorium near the fire station, serving sandwiches, cookies and beverages to hundreds of volunteers. Local businesses donated bread, milk and coffee. "The great common effort is sustained by one hope," the Examiner reported, "to find little Patricia alive."

The searchers fanned out in the Coyote Hills to the north and across the McNally ranch to the south. Four airplanes skimmed low over the expansive farmlands and the orange groves that would be the site for Disneyland three years later. The search continued into Sunday night as volunteers combed vacant lots near town under portable lights from the local fire department.

Authorities were prepared for the worst. The presence of a sharp knife and a large quantity of fresh blood at the crime scene led some to believe that Patty had been dismembered. With five hours unaccounted for, a murderer would have had plenty of time to scatter the remains over a wide area of Orange County. Finding traces of mud, weeds and sheep's wool on the borrowed car, investigators formed special task forces to search the coastal wetlands near Bolsa Chica and the mountainous area near the jail work camp where McCracken had been held.

Other searchers held out hope that the terrified child might have escaped and could be hiding. Deputies drove through the streets with bullhorns, urging residents to "look in all garbage cans, boxes, barns, chicken coops, under houses, small hiding places—and incinerators."

The lost girl's father, Leonard Dale Hull, a 30-year-old construction foreman, stayed at the command post until 10 p.m. The strain was too much for Patty's 28-year-old mother. Neighbors took her home. Only a few days before Patty disappeared, Terry Hull had warned her little girl about the danger of speaking to strangers.

"I'm just numb," she kept repeating. "Just numb."

***

By Monday morning, the story was a front-page headline in every Southland newspaper. Emotions ran high as McCracken made his first court appearance.

"I'll kill him, I'll kill him," Patty's uncle screamed, lunging at the suspect on the courthouse steps. Bystanders tried to restrain Jack Hull as he fought with deputies. Then he collapsed, weeping, and was carried to a first aid area. McCracken showed no emotion in court and gave interrogators at the jail no information about what might have happened to the girl.

The Hulls kept a vigil at their home as the search continued. Patty's brothers watched a crowd of newsmen gather on the front lawn. Their mom kept the boys' hair slicked back because of all the "company that keeps coming" to take pictures of the family huddled on the living room couch.

Hundreds of boy scouts and 1,000 Marines joined the search. Two helicopters and four more planes searched from the sky during all daylight hours. One plane crashed. No one was seriously injured in that mishap, but another volunteer searcher was killed in a car accident on the way to the command post. Monday and Tuesday passed with no sign of Patty.

Investigators collected more evidence. The blood in the cabin did not match the suspect's blood type. Small fingerprints pressed into the window of the suspect's borrowed car were compared with the prints on Patty's piggybank. A white smudge on the back seat upholstery of the car appeared to have been made by polish from Patty's saddle shoes. The couple who witnessed the girl's apparent abduction from the movie theater had picked McCracken out of a lineup. So did the other children who claimed he had "bothered" them in the theater. Mary Plonske, a 74-year-old woman who rented a room in the same motel court, came forward to say that she saw the bartender's car outside McCracken's cabin after 9 p.m. Searchers were instructed to look for a tasseled yellow bedspread that was missing from McCracken's room. The district attorney told reporters, "Even if we don't find the little girl's body, I believe we have enough evidence right now to convict McCracken—and get the death penalty."

***

The ongoing search for the missing child was grist for the growing rivalry between the two leading TV news stations in Los Angeles. In 1951, it was the nation's most competitive television market with four independent stations and, with the CBS purchase of Don Lee's KTSL, there were now three stations owned by networks. The network-owned stations aired scheduled newscasts from their studios but didn't break into the programming supplied on kinescope from the East Coast. No one could imagine pre-empting Milton Berle's popular comedy show for even the most dramatic local news story. The other independents, KHJ and KLAC, usually were more interested in carrying sports and other scheduled events. That left Paramount's KTLA and KTTV, the Los Angeles Times-owned station that had lost its network affiliation with CBS. These two fiercely competitive, independent stations saw local news remotes as a way to establish an identity and attract more viewers.

On Tuesday night, KTTV Channel 11 began live coverage of the search for Patty Jean Hull. Channel 5's Landsberg hated getting beaten by the competing station. Bill Welsh, who had become a household name during KTLA's coverage of the Kathy Fiscus story, had defected to KTTV and was on the scene.

It wasn't that KTLA lacked the technical facilities for a live telecast from Orange County. That part was becoming a routine. For Landsberg, it was a matter of taste. He refused to air advertisements for feminine hygiene products and mortuaries. He banned low-cut costumes for chesty female entertainers who appeared on his station. He believed television should be a welcome guest in the home and not offend families watching together. His heart ached for Patty Jean Hull's family, but he worried that the story of a young girl victimized by a sexual predator crossed the line of acceptable content.

"We were always protective of our news audiences in the 50s," reporter Stan Chambers would later write. "We never used the word 'rape.' We said 'assault.' We never talked about venereal disease, prostitution or personal behavior. We never talked about abortion, homosexuals or people living together. We never said 'hell' or 'damn' on the air."[64]

But Southern California residents were outraged over the details of the Patty Jean Hull case. How could a known sexual predator be allowed to prowl the streets and movie theaters, apparently free to molest children at will? A Los Angeles County supervisor called for tougher laws to require tighter police supervision and even sterilization of known sex offenders. That was the justification Landsberg needed to dispatch a remote camera to the search headquarters at the Buena Park fire station. Landsberg admonished his crew to treat the story "not as a news beat, but as the human interest story of a great tragedy."[65]

On Wednesday morning, KTTV again was on the air first. KTLA's live coverage began shortly after at 8:30 a.m. with newsman Stan Chambers joining Dick Lane at the microphone. Landsberg stuck to his preference for selecting on-air talent based on ability to ad-lib rather than journalistic credentials.

Lane was a movie actor and a popular announcer for professional wrestling matches on KTLA, which starred such colorful characters as "Gorgeous George" and were syndicated nationwide on kinescope.

Lane and Chambers appealed for additional jeeps, diving equipment and other items needed by searchers. The KTLA switchboard lit up with callers offering donations and advice. Parents and civic leaders called to demand that known sex offenders be locked up. KTTV's cameras got the first scoop by being at the jailhouse to bring viewers a glimpse of the suspect being moved from his cell. To keep up, KTLA quickly stationed a second camera at the jail. KTTV's Bill Welsh scored another coup by interviewing search leaders in their map room. But the most dramatic moment clearly belonged to KTLA as Dick Lane conducted the first television interview with Patty's father.

Leonard Hull said he believed his daughter was "better off now than the rest of us because she's probably dead and out of her misery." Viewers shared the heartbreak as the six-foot tall construction worker and the TV wrestling announcer broke down and sobbed together. Lane excused himself and turned the microphone over to Stan Chambers as he struggled to regain his composure. Hal Humphrey, TV editor for the Los Angeles Mirror, wrote, "Regardless of how one feels about televising such a scene, it was impossible to deny the terrific impact of that interview."[66] Both stations stayed on the air during the night and replayed edited film transcriptions of the day's events for those unable to watch earlier. All this *before* the next day's paper hit the streets.

Inventing TV News

Print reporters were scooped again. The Hollywood Reporter noted that "Los Angeles newspapers ran head on into TV as a rough news competitor... Chief beef of the newsmen was that police and other officials were too willing to talk to KTTV's Bill Welsh and KTLA's Dick Lane and Stan Chambers, consistently giving the video men news beats on the strength of having 'known' them through their steady appearances on TV as name personalities."[67]

The dueling telecasts continued on Thursday morning. Joining Chambers at the KTLA microphone was newsman Dick Garton, replacing the emotionally spent Dick Lane. Rumors circulated that a body had been found. "In line with KTLA's policy of truthful reporting, these (unconfirmed) reports were given as such rather than as facts," a station publicist would claim. "Later KTLA's cameras were exclusively on the police officer who officially confirmed the report and called the search to an end... A KTLA film camera crew was dispensed to Trabuco Canyon and filmed the activities there for showing on KTLA later in the day."[68]

What the KTLA press release implied but didn't say was that the competing coverage on KTTV was much more graphic, showing still photos and film of the child's sheet-covered corpse less than four hours after the grim discovery. It was a shocking image for the television audience in the 1950s.

***

Charlie Martinez had an eerie feeling. The 45-year-old mine supervisor remembered seeing a parked car in the remote Live Oak area of Trabuco Canyon on the Saturday night that the little girl from Buena Park had disappeared. At the time he thought it might be a couple of teenaged lovers who wanted to be alone. But after hearing about the frustrating search for the missing girl, he finally decided to tell someone. His wife called a neighbor, who called Neil Parker, a Forest Service ranger. Parker and his boss, Joe Scherman, raced to the spot that Martinez remembered. They found tire tracks. On the far side of a barbed wire fence were footprints that led to a bloody towel thrown over a bush. A broken branch seemed to mark a lone, shallow grave. The two rangers dropped to their hands and knees and tore into the loamy soil with their bare hands. Scherman grasped something and pulled up a child's two-tone saddle shoe.

Patty Jean Hull's body was hidden 33 miles away from her home, deep in the Santa Ana Mountains. Her battered head had sustained some 15 blows, perhaps from an axe, the coroner said. Her small fingers were clenched as if grasping something. The body appeared to have been washed after she died. Further testing would be needed to determine if she had been sexually assaulted. The bloodstained yellow bedspread from McCracken's cabin was found in another shallow hole nearby. The suspect's mother confirmed that it belonged to him. She also said her son "had the mind of a six-year-old."

Investigators rushed McCracken to the gravesite, hoping to elicit a confession. District Attorney James Davis was blunt. "After you buried this little girl in this grave did you kneel and say a prayer for her?" he asked the suspect.

McCracken remained unmoved. "My attorney told me not to say anything."

"Is this yours?" Davis held up the bloody bedspread.

"My attorney told me not to talk."

The frustrated prosecutor signaled to the news cameramen who rushed over to photograph and film the silent suspect beside the grave. Investigators also took McCracken to the morgue to view Patty's body. A court stenographer's pen was poised to take down a confession.

"I don't intend to talk," McCracken said.

The next problem was how to return the suspect to his cell. A crowd of about 500 had gathered outside the jail, jostling for a glimpse of McCracken. The onlookers were demanding justice in full view of the live TV cameras.

"We thought there was going to be a riot," Stan Chambers recalled. [69] KTLA interviewed one man who insisted that McCracken be lynched. One of the announcers firmly reminded the protestor that "due process of law is the pattern of justice in the United States." Deputies managed to keep the situation under control as the TV newsmen for both stations signed off, pledging to return with live coverage of the next day's grand jury proceedings. As promised, live coverage continued from an anteroom in the courthouse where Henry Ford McCracken was indicted on charges of murder, kidnapping and child stealing. The judge ordered him held without bail.

***

The Patty Jean Hull murder was the first live telecast of an ongoing crime investigation. Variety reported that "it gripped Southland viewers... as no event has since the Kathy Fiscus tragedy over two years ago... and was seen by many more viewers." KTLA and KTTV each broadcast more than 31 hours of coverage over three days, drawing mixed reviews and igniting a debate over whether the telecasts had any public service value.

Variety noted, "Two stations were taking different approaches on the killing with KTTV emphasizing the sensational, KTLA playing it more conservatively, factually."[70] On Friday, as McCracken was being indicted, Mirror columnist Humphrey observed, "It is like holding that proverbial bear by the tail. It's tough to know when to let loose. From the standpoint of interviews and showing the scenes involved in the tragedy, both stations had exhausted the story by late (Thursday) afternoon. How long they could remain telecasting and still hold the viewers' interest was proving to be a tough decision for the station executives."[71]

The Los Angeles Times praised its own station, KTTV, for being first with the gruesome story. KTLA general manager Klaus Landsberg defended his station's more restrained coverage. "This is a democracy," he stated in a press release. "We are bringing public opinion in its strongest form to TV audiences with concrete suggestions from the people most strongly affected by this tragedy and crime. We're bringing it to millions of viewers as straight information, on the basis of which they can make up their own minds as to what type of legislation and action from the authorities the situation may demand."[72]

While some community groups demanded tougher punishment for child molesters, the prosecutor in the Hull case saw a different problem. "What California needs is jurors with sufficient courage and backbone to enforce our present laws," said District Attorney James Davis. "Life imprisonment is provided for sex offenders but jurors often do not believe statements by small children who tell of being molested."

A Los Angeles parent-teacher organization commended KTLA for performing a public service: "With this type of factual reporting, displaying the combined efforts of sincere people, television is fulfilling a human need in this area... We appreciate the unsolicited comments condemning the use of Saturday motion pictures for 'baby sitters.' We feel that a great good can come in the form of concerted action by informed citizens who will appreciate the seriousness of such a situation."

A state assemblyman from Los Angeles proposed a bill to set aside sections of movie theaters and make it mandatory for unsupervised children to sit in the reserved seats, away from strangers. One mother proposed her own solution in an interview on KTLA, creating an ironic conflict of interest for the only TV station owned by a movie studio. "I'll never let my children attend any more movies on Saturday afternoons," she declared. "I'll keep them at home where they can watch television right under my eye."[73]

The case continued to draw national attention. Henry Ford McCracken testified in court that he had dreamed of a tobacco-chewing dog telling him to kill his landlady. He said that he had "blacked out" during the crime with no recollection of the little girl. The first trial ended in a hung jury. On September 19, 1951, a second jury found McCracken guilty of murder. After a losing a long court battle to be declared criminally insane, McCracken was put to death in California's gas chamber on February 19, 1954.

***

As KTLA and KTTV cameras brought disturbing reality into the living room, a Paramount movie was showing the potential dark side of the media competition. Billy Wilder's 1951 film "Ace in the Hole," starred Kirk Douglas as an unethical newspaperman. Although many believed it was inspired by the Kathy Fiscus story, the plot more closely resembled the attempted rescue of Floyd Collins from a Kentucky cave, an early test of live radio news coverage in 1925. The headline-hungry reporter in the film tells a man trapped underground that he could become more famous than Floyd Collins if he survives. Then, while the trapped man is dying, the reporter delays the rescue effort and distorts the story to maximize public interest and further his own career. TV camera crews and radio newsmen with microphones arrive as the scene literally becomes a carnival, implying that broadcasters were as culpable as the tabloid newspapers when it came to exploiting a tragedy to draw an audience.

And the audience for TV news was growing. As early as 1947 a survey by Television Research Inc. indicated that 46 percent of those who owned television sets were going to the movies less often as a result. Another sample in 1950 showed three out of four families with television went to the movies less frequently, a trend that continued for several years.[74] "Ace in the Hole" didn't draw many viewers away from their TVs, even after it was re-released as "The Big Carnival." The film was a critical success but a box office disappointment, despite its prophetic message.

Suburban families in post-war Southern California had experienced a loss of innocence as a result of the Patty Jean Hull tragedy. So had local television news.

Chapter Three: The Cold War Hits Home

> *We look with so much humility upon the awesome force that was created by man.*
>
> —*Grant Holcomb, TV newsman describing the first live broadcast of an atomic bomb detonation, April 22, 1952*

The outbreak of the Korean War in 1950 intensified the public's need for news. All seven Los Angeles TV stations added more news programs to their schedules.[75] Independent KLAC hired a globe-trotting foreign correspondent, Clete Roberts, for nightly war commentaries. Southern California residents wanted to keep track of loved ones fighting overseas. KTTV's newsreel crew chronicled Long Beach reservists on a mission over "Red Korea."[76] KTLA provided live coverage of the Marines leaving for active duty and televised their return to Point Loma, near San Diego, as the conflict was winding down. Emotional pictures of servicemen being reunited with loved ones needed little narration. The Los Angeles Mirror commended the KTLA announcers for being "smart enough to let the viewer's own emotion react to what they saw, without giving any verbal prompting."

"This broadcast was further proof of Klaus Landsberg's understanding of the television audience," Stan Chambers would later write. "He let them share in the great emotional moments of the time. He involved them in the spectacle and let them experience the moment as it was happening. The telecast reflected the feelings of an entire nation as the war neared its end." It also showed that KTLA could bring in a live signal originating more than 100 miles from Mount Wilson. [77]

This know-how came in handy when KTLA had to compete with the networks on big national and international stories. Landsberg threatened to set up his own microwave link to San Francisco, forcing NBC to permit pool coverage of General MacArthur's return to the United States, as well as the Japanese Peace Treaty signing in September 1951. But the conference was also the first West Coast event transmitted live to the East on the newly installed transcontinental relay. The following month, the World Series would be transmitted from East to West. It was television's version of the railroad's Golden Spike, uniting the country from coast to coast.

CBS and NBC were now building TV production facilities in Hollywood. Network programming posed a threat to independent stations, especially KTLA. In 1948, the FCC ruled that DuMont and Paramount together owned more than the legal limit of five stations, dashing hopes for a fourth network. For independent stations in Los Angeles, news coverage became more important than ever.

In September 1951, the House Un-American Activities Committee held hearings in Hollywood. The 1947 hearings that led to the blacklisting and jailing of the Hollywood Ten had been held in Washington. As the hunt for alleged communist sympathizers moved even closer to home, KTLA and KTTV jockeyed with ABC-owned KECA to cover the proceedings live.[78] Countless film industry careers would be ruined as some witnesses named names and described a communist plot to infiltrate the motion picture business. Most witnesses invoked the Fifth Amendment and refused to testify. The first day of coverage on KTTV was limited to an audio feed of the proceedings with cameras outside.[79]

KTLA's Landsberg scoffed, "I feel under those circumstances it wasn't TV coverage and we're not going to become a radio station." The next day the committee reversed itself and allowed a pool camera into the flag-draped hearing room, with KTTV, KTLA and ABC-owned KECA each claiming credit for the decision. "Now television can do a real job," Landsberg said. "Its impact is there by showing the character of the witnesses while they're testifying."[80]

KTTV's pool feed, carried by four stations, was commended as "a great service to the community in general but a great factor in the education of the public in the way in which our government works."[81]

***

Experts who had predicted the quick demise of independent TV were confounded when local programs on KTLA continued to dominate Los Angeles ratings for several years after the inauguration of the network coast-to-coast relay. In addition to featuring popular bandleaders like Spade Cooley and Lawrence Welk, the station also boasted an award-winning schedule of public service shows and its signature spot-news remotes.

According to a 1951 trade magazine article, "KTLA is far, far ahead, audience-wise, of its closest competitor in the Los Angeles area; it also carries by far the most extensive schedule of community service programs of any of the seven stations. KTLA's donation of cash, time, facilities and personnel to the people of Los Angeles averages more than $1,000 per day... also the station carries practically all the special (news) events of importance which are available to television."[82] The writer calculated that KTLA spent $400,000 to $500,000 per year on unsponsored live news coverage and public service programs. All stations lost money on news and considered it "the price of the license."[83] But it was also a way to build a station's audience and identity.

"Local news coverage isn't important on film," Landsberg believed. "It's important if it's live because, again, you are taking people there. You are really furnishing a window; but if it's canned it shows up as canned on television. And you can't get away from it. The audience does not want it; they don't like it." Departing from the standard practice of shooting local events on film for later broadcast, Landsberg declared, "... your mobile unit, even if you have only one, is ten times as valuable as three film news services."[84]

Of course, Landsberg couldn't count on big news happening every day. He had to find some way to be live even when nothing was going on. He developed "City at Night," a program that foreshadowed the "live for the sake of live" reports that would be common in local TV news decades later.

"City at Night" each week originated from a location that was kept secret until airtime. Viewers had to tune in to find out if the destination was a factory, nightclub, hotel, tourist attraction, special event or street corner. Although not billed as a news program, Landsberg took care to preserve its editorial integrity. He rejected a sponsorship offer from Richfield Oil because the company wanted to choose the locations. Variety reported an unusual contract with Santa Fe Railroad, allowing that the commercials "may be clipped or eliminated if the occasion arises. (Sponsor) will have no say in what is covered on the show."[85]

"City at Night" required the element of surprise because Landsberg feared the more serious topics might drive viewers away. "We cover any type of activity that takes place at night. We never announce where we're going to go... If I told you that tomorrow night... we were going to Douglas Aircraft to show the manufacture of airplanes there would be a good number of people who would say, 'Gosh, I don't want to see the factories. Why?' So we don't tell them where we're going. Once you have the audience, they're so fascinated they won't turn away."

The notion of a live encounter with something 'real' promised what broadcast historian Mark J. Williams has called "a decidedly new, pragmatic and 'untainted' vision of the world, important qualities to promote in a climate of post-war prosperity and Cold War paranoia."[86] Live coverage of congressional hearings and troop departures allowed viewers to judge the confusing world for themselves.

Klaus Landsberg was also fighting a personal battle. In 1951 he was diagnosed with advanced skin cancer and was told he would need a series of operations. As get-well wishes poured in from his competitors, he often ran KTLA from a hospital bed.[87] Still, his greatest technical achievement was yet to come.

***

On a chilly February morning in 1951, television crews gathered atop Mount Wilson and pointed their cameras in the direction of the Nevada desert. The Atomic Energy Commission (AEC) was testing an atomic bomb. Although the test site was off limits, Los Angeles stations were going to put an atomic flash on TV. KTTV began broadcasting at 4:30 a.m. KTLA went on the air 30 minutes later. Each station electronically darkened its picture so there would be no chance of missing the bright light. KTTV's broadcast was silent. Dick Lane provided KTLA's commentary from Mount Wilson while newsman Gil Martyn was stationed on the roof of the Flamingo Hotel in Las Vegas to provide commentary via a radio hookup. A film crew from a third independent station, KLAC channel 13, was also in Nevada to shoot film for its evening newscast. An estimated 30,000 TV viewers got up early to witness the blast. It came shortly after sunrise.

"Seconds ticked by and suddenly out of the east a blinding flash erupted on the screens," the Los Angeles Times reported.[88] "The mountains disappeared as if by magic. For what seemed like minutes the television screen was a ball of brilliant fire." KTTV stayed on the air for seven minutes in hopes of recording a sound or a shock wave. KTLA's Gil Martyn interviewed Las Vegas gamblers who were unimpressed. The TV networks had sent only still photographers to cover the test and they had been scooped by the local independent stations. KTLA replayed a kinescope of the blast several times that evening.

The U.S. was no longer the world's only atomic power. Americans knew that the Soviet Union also had detonated a nuclear weapon in 1949. Southern Californians, including the mayor of Los Angeles, feared that the region's aircraft factories and oil refineries would be a major target in the event of a Soviet nuclear attack.[89] The U.S. continued to test its atomic weapons as both print and broadcast newsmen besieged the AEC for permission to observe the detonations at closer range.[90]

Aside from national security considerations, the technical problem with TV coverage was that AEC officials used their own microwave relays to observe the explosions. They didn't want interference from radio and TV stations, which ruled out 75 percent of the equipment licensed by the FCC for use by commercial broadcasters. There was also a financial obstacle. The AEC turned down a request for sponsorship, which meant that any station covering the blast would have to spend money with little or no prospect of recouping the expenses through advertising. But civil defense officials needed public support for atomic warfare programs and knew a televised bomb blast would make an impression on a vast audience. Officials held a March 28 conference call to invite the networks to cover an "open atomic detonation" on April 22, 1952. On a hunch, they also invited an executive from a local, independent station— KTLA's Klaus Landsberg.

When the group met in Chicago on April Fools' Day, the phone company delivered the bad news. It would cost more than $60,000 and take at least six months to set up 10 to 12 relays, covering more than 277 air miles from the Nevada Proving Grounds at Yucca Flats to Los Angeles. Something as tiny as an electric razor could interfere with the blast, so all the equipment would have to be in place and approved at least one day ahead of time. And the bomb test was only three weeks away. The networks were wary. The phone company backed out. "They said it couldn't be done," AEC public relations supervisor Charter Heslep would later recall.

Only Klaus Landsberg was willing to assume the huge risk. If he achieved the seemingly impossible signal, the networks would share the cost of the telecast. If he failed, KTLA would foot the entire bill. "Go ahead and try it," said Paramount executive Paul Raiborn, Landsberg's boss. They named the project "Operation Big Shot."

***

KTLA engineers gathered around Landsberg's kitchen table on a Sunday afternoon. "Klaus pulled out a bunch of maps and we all got down on our hands and knees and looked at different high spots from here to Las Vegas to figure out where we could put our relays and how many relays we would require," KTLA engineer John Polich remembered. [91] Mount San Antonio, above the village of Wrightwood, had a direct line of sight to Mount Wilson and a tiny cabin on top. But there was no road to the 8,500-foot summit. The only way to haul in the equipment was to put on snowshoes and hike through eight feet of snow. The hulk-like Polich, a former star hockey player, strapped some of the gear to his back and led the way. By April 5, a parabolic dish was in place. Engineers spent hours on their walkie-talkies tweaking the signal until it reached the KTLA transmitter, 28 miles away. But that was only one link in the chain with 16 days to go.

Phone lines were ordered for News Nob, where the reporters and cameras could observe the blast from just 11 miles away. But what about the shock wave? No one knew if the light and heat of the blast would blow the picture off the air. So Landsberg installed two cameras. One would have its lens capped until after the blast, just in case the first camera failed. Above the AEC control point on Mount Charleston, 41 miles away, backup cameras would be ready if both close-up cameras went dark. Klaus wasn't concerned that the signal had to travel through two small "saddles," one of them only 150 feet wide. He was confident he could thread that needle. A much bigger problem was closing the remaining gap in the relay between Mount Charleston and Mount San Antonio with 14 days to go.

A 12-mile search turned up no place for a dish on Clarke Mountain, a 10,000-foot peak, 60 miles south of Las Vegas. Winds howled at 80 miles per hour and aerial reconnaissance had to be scrubbed. Landsberg and Ray "Pappy" Moore, KTLA's 55-year-old chief engineer, kept looking. They bounced over the scrubby mountain trails in Landsberg's big, black Chrysler sedan. They were gone for two days. Los Angeles newspapers speculated on whether the telecast was doomed. The exploits of the German-born engineer and his television

Inventing TV News

team were the talk of the 3,000 scientists and military personnel at the test site, Camp Mercury. Is he going to make it? Is he out of his mind? Just as the AEC was about to send out a search party, Landsberg checked in with the good news. He had discovered "Mount X."

The uncharted mountain had the perfect shelf for a dish at 6,500 feet. But the only way to reach it was by helicopter. Landsberg called the El Toro Marine base. They had two giant Sikorsky HR-1 helicopters, but they had never flown above 5,000 feet. However, the idea of taking part in the first televised atomic bomb test appealed to the military brass. Soon, the Marine choppers were hauling six tons of television gear up to the isolated mountain shelf. The transmitting dish, eight feet in diameter, wouldn't fit in the hold and had to be tied on. Landsberg and three other TV engineers wore parachutes, just in case something went wrong, as the choppers deposited them on the remote ledge. Four pilots worked in shifts, braving the desert heat, sandstorms and mountaintop blizzards. It took 24 flights. One of them didn't make it. "When we were unloading the equipment on Mount X we smashed up one of the helicopters quite severely and two of our boys got hurt," Polich recalled. "But we all survived." The survivors salvaged the equipment from the downed chopper and hand-carried it back down to the base camp.

The next problem was establishing a signal from Mount X to Mount San Antonio, 140 miles away. The manufacturer guaranteed the transmitters to cover only 40 miles but Landsberg was confident they could do more. The engineers worked all night and established what looked like a strong signal. Just after the sun came up, the signal died. It took two more days to make it work.

On Tuesday night, with just one week to go, Landsberg was ready to line up the final 50-mile link in the relay from the AEC control point at Mount Charleston to another peak called Stone Mountain just south of Las Vegas. It was a clear, crisp night and the lights of the casinos on the strip glittered like distant jewels. The crew on Stone Mountain flashed a powerful beam to establish a line of sight. Landsberg and the crew on Mount Charleston saw nothing. Some kind of obstacle was in the way. But what? The next day a ranger peered at the map and studied the horizon. There was an uncharted ridge blocking the line of sight. They would need two more TV crews to relay the signal through Nellis Air Force base, and they had only four days left. The detours had lengthened the relay to 400 miles, with shivering engineers camped out on the isolated relay peaks. But they now had a plan to transmit a picture all the way to Los Angeles.

***

On Mount Charleston, Pappy Moore could hardly believe his eyes. Yes, he could see Mount X through some kind of natural hole in the massive mountain range. He picked up the phone and called Landsberg. This lucky break seemed too good to be true because it meant the Stone Mountain and Nellis links would not be needed, shortening the signal to less than 300 miles. That night they conducted another flashing light test and found out that Pappy's vision was no mirage. All they needed to do was move the Mount X receiving dish over to the next hump on the mountain and patch it into the relay with coaxial cable. A good quality picture from the Charleston site was now reaching the KTLA studios. Landsberg relaxed and everyone went into Las Vegas for a little fun. The celebration was premature.

A vicious sandstorm whipped up the testing ground at Yucca Flat. Mount Charleston and Mount San Antonio were hit by blizzards, burying the transmission equipment under a foot of snow. The entire relay system was useless with just two full days to go before the blast. The engineers worked all night to get it up and running again for the AEC's frequency check the next day. If any of the TV frequencies interfered with the testing, the broadcast would be cancelled. It was also important to know if the AEC's transmissions would mess up the TV picture. At 10:35 a.m., the AEC gave the clearance Landsberg needed to hear: "Okay on television."

On the eve of the detonation, KTLA tested the relay by putting on a pre-bomb telecast featuring Tony Martin, Vaughn Monroe and other acts that happened to be playing in Vegas. Los Angeles viewers commented on the high quality of the picture.

Everything looked fine until the big day. The KTLA crew arrived at 1 a.m. for the final setup on News Nob, only to find that their press passes weren't valid until 7 a.m. The AEC's public relations man raced over in his pajamas to vouch for the nine exhausted and unshaven men. KTLA planned to go on the air at 8:45 a.m. with the networks picking up the signal at 9 a.m. It appeared Klaus had done the impossible.

Or had he? AEC procedures called for a radio blackout seconds before the test, so Landsberg's mobile unit lost communication with Pappy Moore at the AEC control point on Mount Charleston. At 9:20 a.m., the generator on News Nob blew its thermostat, cutting all power to the close-up cameras. The shot went dark just as the announcers were beginning their final countdown to the scheduled 9:30 blast.

"What happened?" Klaus screamed. "Where's the picture? Somebody out there, fix it!" They had 10 seconds, not enough time to restore power and warm up the cameras again.

Forty-one miles away on Mount Charleston, cameraman Robin Clark was counting down the seconds from the backup camera. He saw an Air Force bomber in the distance and focused his shot, wondering if he could see the actual drop. He had no communication with anyone, no way of knowing if his picture was even on the air. He didn't find out until later that he had the *only* picture on the air.

***

The broadcast began with Clark's long shot of a barren "sandy stretch" of desert, with announcers Grant Holcomb and Fred Henry in casual hiking clothes reporting from the location known as News Nob. The reporters believed they were close enough to experience the terrifying spectacle of the bomb but far enough away to avoid any real danger. The KTLA announcers filled time as they scanned the sky for three B-50 aircraft, one of them carrying a "nominal size" bomb.

Inventing TV News

"Bombs away," said Holcomb said as he began counting down the seconds to the blast. A distant voice could be heard over a loudspeaker.

"Five, four, three..."

Robin Clark pulled back to a wide shot and centered on the area where he thought the bomb would fall. He was just in time. The brilliant flash of the exploding bomb overexposed and blacked out the entire screen. A ball of white appeared near the horizon, peering out from the black circle surrounding it. The announcers fell silent. The loudspeaker warned that the shock wave would arrive in 30 seconds.

"Well, there it is," Holcomb declared. "The first public demonstration, and the biggest continental atomic detonation in the history of the world." The loud boom of the shock wave punctuated his sentence. It was followed by a blast of heat that the newsmen would compare to "opening the door to a furnace, and you felt yourself slapped in the face of that heat wave, a combination shove and slap." The viewers at home saw the fuzzy black and white unfolding image of a cloud, which appeared orange, brown and red to the newsmen on the scene.

"Fred," said Holcomb, "just looking away from that beautiful mushroom that's turning into that fabulous white now, and down to that ugly gray all across the floor of this valley, you can't help but realize that you could put—right inside this proving ground from where we are—the entire island of Manhattan, New York City. And if you look at that dirty, ugly gray base, you see what that particular weapon can do."

Ominously, he added, "...the enemy certainly has similar weapons."

At this point, the News Nob generator was up and running and the "close-up" camera had warmed up again. Landsberg was finally able to cut to the shot. To the viewers at home, it appeared that KTLA had intentionally opened with the wide-shot camera to show how the atomic blast dwarfed everything around it.

The close-up view from News Nob showed technicians arriving with Geiger counters to detect the amount of radiation at the fringe of ground zero. The newsmen knew there were troops in the area, but the military would not confirm rumors that 1500 servicemen were in foxholes less than eight miles from the blast. Planes and trucks brought in 120 paratroopers from the Army's 82nd airborne division at Fort Bragg, North Carolina. "When it is safe," said Henry, "the paratroopers will drop in, in back of the bomb, the idea being for the troops to join up and take the theoretical objective."

Viewers saw a fire on the horizon, one of the three blazes ignited by the tremendous heat of the bomb. The next scene showed the slender Holcomb atop News Nob, surveying the test site with Hugh Baillie, a stocky, crew-cut correspondent for United Press.

"Still a lot of dirty dust out there, isn't there, Hugh?" Holcomb asked.

"Yes, I've seen a lot of bombardments," replied the experienced war correspondent, "but nothing like that... It was, of course, one of the most tremendous things the eye of man could witness." Baillie surveyed the smoke plumes over the desert and imagined the scene if the bomb had been dropped on a city. "It would be devastated, filled with dead people. Not many dying people, most of them would be dead, I expect."

The two newsmen then gave viewers a behind-the-scenes look at the reporters covering the story. Baillie pointed to the GIs who were manning the teletype machines labeled with the names of the various wire services. Holcomb pointed out a KTLA camera on top of a car that was precariously parked on Monkey Rock. Also shown was a helicopter that helped to bring in some of the equipment. The atomic cloud was still visible in the distance.

"It now looks as innocent as any other cloud in the sky, doesn't it?" observed Baillie. "You can't tell it from the other clouds. Even if you're a scientist... All those strange shades of orange and red and the glow is all gone, so it floats along quite harmlessly. I wonder what would happen if somebody flew an airplane through that by mistake?" Holcomb chuckled in response, then turned serious.

There was an awkward pause. Had they goofed? The AEC had downplayed the possibility of any danger from the radioactive fallout and wanted newsmen to do the same. "I do know that they track that cloud with B-29's all over the United States until it dissipates itself," Holcomb pointed out.

A general arrived at News Nob by helicopter and was respectfully welcomed by the press. Planes carrying the airborne troops took off and headed for the jump site. The home audience saw fuzzy pictures of the paratroopers as they parachuted over the test area. For background and context, the TV announcers turned to another print reporter.

William Laurence of the New York Times was the only newsman who saw the first atom bomb detonated at Los Alamos, New Mexico on July 16, 1945.

"My reaction, Grant," he said, "is really that you never see a second atomic blast. You always think it's a first. No matter, I think, how many you see you always have a new experience, as though you've never seen it before."

What the Times reporter didn't say was that, in a real attack, no one in the target area survives to see a second atomic bomb. But Los Angeles viewers would live to see a second nuclear blast from the safety of their living rooms. Gil Martyn and Stan Chambers broadcast live from Mount Charleston when another bomb was detonated the following week. But the irony was lost on Laurence, who went on to describe how the bomb generates a force equivalent to 20,000 tons of TNT from as little as one gram of uranium or plutonium.

"Would you say they split a little bit of the universe this morning?" Holcomb asked. Trying to answer, the newspaperman fumbled for the words to make complicated nuclear science sound simple enough for TV.

"It was at least 100 suns, a million degrees Fahrenheit," Laurence continued, adding that the radioactivity dispersed at a high altitude and didn't hurt anyone.

Fred Henry quickly brought the historic broadcast to a close.

***

In April 1952, Broadcasting magazine reported more than 1.2 million television sets in Los Angeles. The first live telecast of an atomic bomb detonation drew a huge audience but mixed reviews. Mysteriously, some TV sets brought in clear pictures while others received interference or blacked out. The result over the transcontinental hookup wasn't much better. "Television fell flat," the Copley News Service reported. "It was as if someone had lighted a match." As Pentagon brass gathered around a monitor, one officer remarked, "I've seen a bigger explosion from my youngster's cap gun."[92]

An Ohio newspaper agreed that movie films of earlier tests provided a better picture but lacked "the physical, momentary presence of the explosion... Here was the atom bomb going off right before your eyes. It was not a historical document. It was a living event."[93] Millions of school children watched the blast on TV in their classrooms.

Some viewers who tuned in late thought the mushroom cloud was some kind of commercial. A Sacramento Bee cartoonist speculated that it might be a plug for a new cereal that went "...snap, crackle, BOOM."[94]

"America has learned to live with the atom bomb," wrote columnist Hal Boyle of the Associated Press. He commented that the public no longer feared the next war would automatically lead to nuclear annihilation, since the Korean conflict had been waged without either side dropping the big one. "But the mere fact that millions of housewives could sit in their living rooms and calmly watch an atom bomb burst shows how much the public has lost its fear of this frightful weapon." The Cold War audience, he theorized, "has become adjusted to an atmosphere of prolonged crisis. You can't remain tense forever."95

The highest praise came from legendary newscaster Walter Cronkite, who was covering the blast for CBS and recognized Landsberg as the man behind the technical achievement. He "seemed to have an immediate grasp of the needs of the news business, as well as vision of the possibilities of live television coverage. He certainly was one of the geniuses of the early days of television. He refused to accept anything as impossible."96

But for another legendary broadcast journalist, the televising of an atomic blast merely validated suspicions that TV might never be a source of serious news. "What seemed to concern television isn't the horror of the atom bomb, but the unique picture it makes," Edward R. Murrow remarked.97

***

KTLA's first atomic bomb telecast has been called "the most prominent example of a local station making a national impact." It transformed KTLA from a station that merely shared its kinescopes with other individual stations to a "local source of 'network live' since the feed was used by all three networks."98 Eventually, Williams notes, "KTLA and the whole of the Los Angeles independent television market was forced into a more subservient role." One KTLA insider put it more bluntly: "We went out with a bang."

As Landsberg was setting up the bomb broadcast in April 1952, the FCC lifted its 43-month freeze on new television station licenses, allowing further growth in the power of the

TV networks. Prior to the lifting of the freeze, there were only 128 stations on the air nationally, capable of reaching only half the nation's population. By 1960 there were nearly 1,000 local stations and most were network affiliates.

Los Angeles was typical of the national trend toward network programming from New York, with local stations originating just enough news and public service programs to keep their licenses. The "Chicago School" of production was shut down when all network programming was moved to New York in 1955. Chicago Mayor Richard Daley demanded an FCC investigation of Chicago stations that cancelled money-losing local news programs to carry more lucrative network entertainment. The FCC ruled in 1962 that there was no reason to revoke any licenses but agreed that the network programmers in New York had insisted on stations dumping local programs for network offerings. Despite that finding, all licenses were renewed with no punitive action taken.[99]

Lacking a network affiliation, independent KTLA continued to emphasize local events, broadcasting the 1955 Rose Parade in color and continuing to cover atomic bomb tests. But Landsberg found himself increasingly isolated. Network executives "have no love for 'the boy wonder' as they derisively refer to him," the Mirror's TV columnist wrote.[100] "Landsberg's personality reflects a strange mixture of superb confidence and frustration. Humility is not one of his virtues. He... is pained deeply at the knowledge that many of his contemporaries dislike him." The discord was compounded by a KTLA broadcast of the popular Hollywood Christmas Parade. Landsberg refused to show floats and stars from competing TV stations, a policy that was later changed because of viewer complaints.

On September 16, 1956, Landsberg lost his battle with cancer. Even those who found him abrasive agreed with Cronkite's description of Landsberg as "one of the first advocates of live television, getting the live camera out where the action was. He lives in all of our hearts." An article in the Orange County Register asserted that "TV now owes its inception in part to the self-taught genius who launched a multi-million west coast industry from a panel truck."[101]

At a time when FCC regulations allowed most stations to get by with a token effort to report the news of the day, the Los Angeles 'school' had demonstrated the power of local coverage and foreshadowed the public's fascination with calamity, crime and live reports. Landsberg's engineering gamble on the atomic bomb tests succeeded in bringing the Cold War into the living room, a tradition his KTLA colleagues would carry on.

***

Viewers of a morning cooking show on KTLA in 1959 were surprised to see a sizzling skillet replaced by a picture of a missile on the horizon and the voice of Stan Chambers. "We interrupt this program to bring you a special live report from Vandenberg Air Force Base," Chambers breathlessly announced. Then, a loud boom. Flames shot out of the tail of the rocket as the first ballistic missile launched from the West Coast headed skyward. For reasons of national security, defense officials would not allow any advance publicity. But live coverage of the launch was another broadcasting first for KTLA.[102]

That same year KTLA put a human face on the Cold War by covering the visit of Soviet premier Nikita Khrushchev to the set of the movie "Can-Can."[103] The local, live telecast was picked up by all the networks. TV pictures showed the Soviet dictator and other dignitaries in the balcony of a studio at 20th Century Fox, observing a racy dance by actress Shirley MacLaine and the chorus. KTLA newsman Bill Stout labored in vain to get anyone to comment on what Mr. and Mrs. Khrushchev thought of the performance.

The wide shot shows a KTLA camera jostling for position with "70 or 80 still photographers." Glamorous chorus dancers provided a visual contrast with the frumpy Russians as the Soviet delegation mingled with Hollywood stars including MacLaine, Frank Sinatra, Louis Jordan and Maurice Chevalier. The bizarre scene didn't provide much in the way of content or context. Only a few months earlier, Vice President Nixon had challenged Khrushchev in the Moscow "kitchen debate" and the two superpowers were racing for supremacy in space by launching dogs and monkeys into orbit. But Stout's heaviest question was "What did he think of the dancing?" and it was never really answered.

Correspondent Clete Roberts picked up the action outside, where the Soviet premier stopped to talk with film star Gary Cooper.[104] Khrushchev "enjoyed himself immensely today, no doubt about it," Roberts declared. "Mrs. Khrushchev was positively radiant and his state of mind certainly equaled hers." What the TV coverage failed to note was that Khrushchev had exploded on the movie set after being told he would not be able to visit Disneyland. "What is it?" the Soviet dictator blustered. "Do you have rocket launching pads there? Is there an epidemic of cholera there or something? Have gangsters taken hold of the place that can destroy me?" Khrushchev's mood failed to improve as he went on to deliver blunt remarks on the possibility of nuclear war.

Roberts noted, "There were some moments during his speech when the friendliness was not there." He reported that the Disneyland visit could not be arranged because Los Angeles County authorities had no jurisdiction over security in neighboring Orange County. The Russian visitor would have to be content with a tour of suburban housing, industrial areas and supermarket complexes in the San Fernando Valley. Roberts then worked his way to the front of the crowd and announced he was shaking hands with Khrushchev. A crowd of newsmen blocked the shot. Roberts then placed his microphone "right at the lips of Madame Khrushchev," who comments as she ducks into a limousine that the movie set performance was "simply beautiful."

"We can't follow the convoy," Roberts said, "because we might lose our cables."

But thanks to another KTLA innovation in Los Angeles, television news would not be tethered to the ground for long.

Chapter Four: Pictures From the Air

Come out of your corners shooting, boys.

—Barney Glazer, TV columnist, observing the cut-throat competition among local news crews in Los Angeles, 1953.[105]

John Silva, the chief engineer for KTLA, worried as he drove to work. The competition was escalating out of control. Whenever there was breaking news in Los Angeles, KTLA and KTTV would lead the race to see who could put the story on the air first. The newspapers were keeping score. And it was getting dangerous out there. Local TV crews in the early 1950s risked their lives to get the most dramatic live pictures.

While covering a dam collapse in Santa Monica, KTLA placed a camera on the roof of a four-story building to get a better overhead shot. According to a KTLA press release, "The building gave way and collapsed. The cameraman survived by jumping into a fire net. The equipment was lost."[106]

KTLA cameras were in harm's way again during live coverage of high surf pounding the coastline at Redondo Beach on January 14, 1953. Buildings from San Diego to Santa Barbara sustained an estimated $15 million in damage, with one cottage collapsing into the sea.[107] Newspapers reported that a teenager rescued a four-year-old boy who slipped off a seawall into the towering surf. But the most dramatic moment of the televised disaster centered on the fate of a KTLA news camera. Columnist Barney Glazer described the scene for his readers: "Were you watching KTLA's dramatic telecast of the cannonading waves at our beach area when the long arms of King Neptune's law reached out and claimed a $30,000 camera? As the camera went to Davey Jones' locker, the picture on your set faded into blackness. That's what it looked like to Jonah inside that whale's kitchen."[108]

Glazer may have exaggerated the value of the lost camera, compared to other reports placing the loss at $13,000 to $15,000. And there was a loss of official dignity as well. The wave that claimed the camera also drenched the police chief and other city officials who were being interviewed live at the time.[109] A KTLA press release noted, "One huge wave, a blackout, and a lost camera. Once again, the cameraman barely got out in time." But the Hollywood Reporter found humor in the moment, "...including the comedy bit when the engineer's voice boomed at the cameraman: 'What the hell are you doing? You're missing all the shots!'"[110]

The TV audience loved it. A Paramount Television Productions memo quoted a letter from viewer Tom Danson: "This is one of the reasons why KTLA helps raise the mental level and attitude of the TV viewer."[111] The Paramount-owned station took such risks in the face of more aggressive competition. KTTV, owned by the Los Angeles Times, also covered the damaging surf and some viewers felt that its signal came in more clearly. KNBH, the NBC-owned station, was also competing in news with roving film crews and the slogan

"You'll see more on Channel 4." Channel 5 responded with "Every night is a big night on KTLA."[112]

On many of those big nights, Klaus Landsberg's right-hand man was John Silva. Together, they had engineered many TV triumphs. And, occasionally, they felt the sting of being beaten by the competition.

***

Deep in the Tehachapi Mountains, 23 miles south of Bakersfield, the White Wolf fault rumbled to life at 4:52 a.m. on July 21, 1952. The 7.7 magnitude earthquake, comparable in force to the 1906 temblor that devastated San Francisco, left at least 11 people dead and was felt as far away as Reno, Nevada. It also rattled KTLA reporter Stan Chambers' home in Malibu. After making sure his family was safe, Stan headed for the KTLA newsroom. Engineers had located the apparent epicenter on their maps and were trying to determine if they could get a live signal to Los Angeles. The news wasn't good. The small town of Tehachapi was 150 miles away and surrounded by mountains that would block the TV microwave. However, the success of the atom bomb telecasts three months earlier persuaded Klaus Landsberg's team to give it a try. KTLA ordered a phone line from the disaster zone and sent a film crew as backup.

But the KTLA atomic bomb broadcasts had also convinced KNBH of the importance of being first with breaking news. NBC-owned channel 4 was on the air with announcer Eddie King reading damage reports at 6:23 a.m. NBC then cut to its network coverage of the 1952 Democratic National Convention in Chicago. While KTLA's mobile units were driving to the scene, a KNBH newsreel crew was in a chartered plane flying to the disaster area and back to Los Angeles. Channel 4 had film of the destruction on the air at 12:30 p.m., showing 1910-vintage buildings reduced to rubble and victims buried under piles of bricks. A station that didn't even have a live news unit had scooped mighty KTLA. [113] But at least they were still broadcasting. The earthquake had caused a power outage in the Los Angeles basin, which knocked independent station KHJ off the air completely.

By the time KTLA's remote trucks arrived in Tehachapi, other stations were on the scene and scrambling to order phone lines. Too late. The telephone company could install only one broadcast link and KTLA had placed the first order. Landsberg agreed to a compromise that would allow other stations to use the linkup—later.

Landsberg's son, Clevie, had tagged along on the shoot. The 24-hour demands of the job had taken a toll on Landsberg's marriage. Divorced and battling cancer, Klaus's idea of quality time with his son was to take him on location. As an adult, Cleve Landsberg would admit to having been "the worst kind of brat," a prankster capable of causing havoc on a set. One time he got hold of a studio fire hose and soaked the elegant evening gowns of Ina Ray Hutton and her all-girl orchestra. "That kid was a pain in the ass," recalled John Polich, the strapping former athlete who stage-managed many KTLA remotes. To keep the child busy, the crew assigned little Cleve to help out with simple production tasks as they prepared to transmit the first live pictures of the earthquake damage.

Cleve Landsberg would grow up to work in the film industry as a production manager and would never forget his front-row seat for the early days of television news. "There I was, a wide-eyed six-year-old in the back of John Polich's pickup truck feeding cable out of the back as we slowly patrolled the devastated main street, sending those live pictures back to Los Angeles. That's what television was all about. I loved every second of it. And I loved my dad."[114]

There were hundreds of aftershocks. A relatively big one occurred while Stan Chambers was on the air. "I felt a deep rumble and powerful movement under my feet," he recalled. "I had trouble keeping my balance but I knew I had to keep on talking and describing how this latest tremor felt. I have no idea what I said but I was greatly relieved when the shaking stopped..."[115]

That night the KTLA crew slept under the trees in a park. Chambers drifted off to sleep on his cot, proud to be "a source of information that helps people cope in troubled times." But the delay had hurt Channel 5 in the publicity wars. By the time KTLA got its live earthquake coverage on the air, the Los Angeles Times was declaring victory for "the film television lads around the local channels (who are) enjoying a loud chuckle at the expense of exponents of live television."[116]

John Silva recalled, "The whole idea was whoever got there first was the hero and the second station was the bum. The newspapers got hold of the rivalry and really promoted it, always reporting which station got on the air first."[117] Crews in the field did whatever was necessary to win.

"A woman had climbed up on a water tower near Beverly Hills," Silva said, recalling one especially dirty trick. "She had threatened to jump and she was up there for quite some time. We (KTLA) got there first. We started jockeying for position to get a signal and a power source. Then KTTV pulled up. Their remote supervisor was standing in front of their truck. He told me, 'The lady up there says that if she sees any television units she will jump. So, we're getting out of here and suggest you do the same.' So, I called Klaus at the station. He said we should get out of there. We drove back to the station only to find that KTTV was on the air live with the story. They just plain told a lie to get us out of there."

***

Superior technology was KTLA's ace in the hole. By 1957, both Landsberg and his original chief engineer, Ray "Pappy" Moore, were dead of cancer. Silva had taken over the engineering operations and was determined to continue KTLA's tradition of technical leadership. The soft-spoken San Diego native became fascinated with television at the age of 12, tinkered with it through high school, and entered the Massachusetts Institute of Technology as a major in Communication Engineering. Two years later he transferred to Stanford. He received his engineering degree just as World War II broke out. He joined the Navy as an ensign, specializing in radar and landing in the Pacific with the Marines. He was on patrol aboard the U.S.S. Shea when the destroyer command ship was hit by a Baka bomb, a Japanese suicide aircraft. The explosion took a heavy toll in the ship's radar room, where Silva was somehow left standing as the men on either side of him were blown apart. Operating a first aid station, the radar specialist administered plasma and morphine to the wounded. Seventy-eight men died. John Silva was awarded the Purple Heart. His next post, with the Chief of Naval Operations in Washington D.C., put Silva in charge of research development for electronic equipment.[118] This background qualified Silva for a top engineering post at the emerging TV networks in New York, but he turned down a post-war offer to join NBC. He and his Texas-born bride, Brooksie, were homesick for the West. He returned to San Diego and wrote a letter asking Klaus Landsberg for a job. After a three-hour interview in KTLA's Paramount garage, he was hired as a TV engineer in 1946 for $50 a week.

"Klaus had a habit of testing people by flinging his cigarette butts on the floor to see if the other person would stomp them out," Silva recalled. "One day, he flung a cigarette butt at me and I said, 'I'm not your servant. Do it yourself.' And we were friends."

The two broadcasting whiz kids became like brothers as they worked side by side in the TV trenches. The technically inclined Silva would take care of the engineering details while Klaus came up with the ideas for new programs. "Sometimes we did 10 remotes per day," Silva remembered. "Klaus wanted to get as much experience as possible, getting a line of sight, surveying the site. Each time we got better."

After Landsberg's death in 1956, Silva continued the KTLA tradition of engineering firsts. He developed KTLA's Telemobile, a station wagon that could broadcast a live signal while speeding along the freeways at 55 miles per hour. But there was still the problem of getting around buildings, hills and anything else that could block the direct line of sight to the transmitter. And in the congested urban sprawl of Los Angeles, having a fast remote unit was no guarantee of beating rival stations in traffic.

So, on a crisp Southern California morning in 1957, John Silva's worries about the competition produced an idea that would give local television news a new perspective. As Silva headed north on the 101 Hollywood Freeway he could see the transmission towers atop Mount Wilson to the northeast, with their unobstructed 360-degree view of Southern California. That was John Silva's light-bulb moment.

"I was driving to work, wrestling with the idea of how we can beat the competition, build a better mousetrap," he recalled. "And I thought to myself, 'Of course! We could do the news from a helicopter.'"

***

Silva kept his idea to himself as he sketched some rough plans. The concept of airborne television was not new and it had originated in Los Angeles. In 1932, engineer Harry Lubcke of experimental station W6XAO beamed up the first in-flight 'movie' to a group of reporters circling Southern California in a Western Air Express tri-engine Fokker airplane. The silent signal on a cathode-ray receiver proved to skeptics that it wasn't necessary to be hooked up to a cable to see television, as long as nothing was blocking the line of sight. But could the signal travel the other way, from the airplane to a receiver on the ground?

The British claimed credit for the first broadcast from an airplane in 1936. Scottish inventor John Logie Baird placed one of his spinning-disk mechanical television cameras on a Royal Dutch Airlines flight. He transmitted pictures of some famous passengers—actors Charles Laughton and Paul Robeson—from 4,000 feet to London below. The primitive equipment could scan only 48 lines per screen, so the picture quality on the three-inch screen was poor. But Baird's attempt preceded RCA's first airplane remote by about three years.[119]

KTLA broadcast the first live picture from a helicopter during a Fourth of July celebration on the aircraft carrier U.S.S. Valley Forge in 1953. Eager to accomplish the first live telecast from the deck of a ship, the KTLA crew tried to drive the transmission truck across Santa Monica beach to meet a Navy landing barge. Beachgoers laughed when the remote unit got buried up to its hubcaps in the sand, but then pitched in to help push. A final tug from the Navy barge's powerful hoist finally freed the truck for the short ride to the carrier anchored off the California coast.

Klaus Landsberg watched a helicopter land on the flight deck. That gave him an idea. He asked if he could put a camera up in a chopper.

"I don't know," an officer answered. "I thought those cameras were too big for a chopper. And... what about the cable?"

Klaus suggested it might work if the cable was long enough. "If the helicopter went up slowly it could go higher than the bridge of the carrier. It would give us a great shot of the action."

The idea traveled up the chain of command. The captain gave the green light for cameraman Ed Resnick to squeeze a huge studio camera into the bubble of a helicopter. Resnick balanced the camera on his lap and got a dramatic shot through the open door as the chopper carefully rose above the ship. Stan Chambers recalled, "It was quite a sight in the darkening sky, hovering over the flight deck with 50 feet of cable hanging out the side door and another 100 feet or so coiled on the deck below."[120]

At the time, Silva remembered, no one imagined that the camera and the transmitting equipment in the truck would ever be small and lightweight enough to dispense with the cable and turn a helicopter into a flying TV station.

But in 1940, NBC program manager Thomas Hutchinson had peered 20 years into the future and envisioned a 1960 control room covering a transatlantic speed-record flight. "A young man... is switching cameras back and forth from the jet plane to the helicopter while a commentator gives a first-hand verbal description of what is happening," he predicted. Hutchinson foresaw a local station in a metropolitan area with numerous air and ground mobile units that could break into a program schedule at any time to provide coverage to its viewers.[121]

***

After several months of working alone, John Silva had a preliminary design and was ready to make Hutchinson's dream of airborne, live news coverage a reality three years ahead of the predicted schedule. He just needed approval from his new boss.

Paramount had picked Lew Arnold to succeed Landsberg as general manager of KTLA. The former DuMont executive was aware that he had big shoes to fill. And he didn't want to risk a failure after only one year in the job. Arnold said no to the TV helicopter scheme. He saw it as a risky investment in something that might not work, tarnishing KTLA's legacy of innovation during the Landsberg years. Silva left the general manager's office with a heavy heart. This might have been just another sad case of a good idea squelched by a bad boss, except that Arnold was fired a few months later.

"The new fellow (they) brought in was Jim Schulke," Silva recalled. "I liked him right off the bat. After I got to know him a little bit I went in and told him what I had been working on. I told him I didn't have very good results with Arnold."

Schulke's reaction was different. "John," he exclaimed, "that's a hell of an idea. What the hell are you waiting for? Move ahead with my approval. Go!" Silva walked out of the office happy and ready to meet the challenge. But Schulke had a few more instructions for his chief engineer: "Choose a select, trusty crew. You must not say a word. The people across the street or the networks, they could literally take it away from us if they liked the idea and thought they could do it themselves."

Terry Anzur

By 1958, KTLA had moved from its cramped quarters near Paramount studios to the sprawling old Warner Brothers lot at the corner of Sunset Boulevard and Van Ness. KTTV, which had been its main competitor in news for nearly a decade, was literally across the street. If KTLA parked a helicopter on the new lot, the rival station's engineers could easily spy over the wall and steal the idea. Network stations with bigger budgets might be able to develop the idea more quickly if they found out about it.

There was also the question of who would risk his life in the initial test flights. "I decided that I would be the one flying while it was developed," Silva said. "Then the unions would get involved and someone else could take over."

Silva picked a trustworthy engineer named Harold Morby to be trained as a helicopter cameraman and engineer. He selected soundman Roy White to work on the audio gear. Both men were sworn to silence and the Paramount machine shop joined the secret development plan. All parts for the helicopter would have to be built behind the guarded studio gates at Paramount or in a secret, locked workshop at KTLA. Only Silva, White and Morby had keys. Stan Chambers called the project "one of the best kept secrets in early television."

Silva had several major problems to solve. In order to meet flight requirements, he had to reduce one ton of television equipment to less than 400 pounds and distribute the weight evenly. Even at that weight, the helicopter could only carry half its normal fuel supply. It wouldn't be able to take long trips but would have great coverage over the city. Silva had to minimize the noise, heat and vibration to a level that would not interfere with the quality of the picture. And he needed a flying antenna capable of directing its signal to the Mount Wilson transmitter continuously from any angle.

"Silva designed his helicopter at the right time," Chambers recalled. "Transistors were making miniaturization possible. He was able to incorporate all the new science into his transmitting gear and cameras."[122]

"I had it pretty well engineered out," Silva explained. In a memo to Schulke he wrote, "There is positively no doubt in my mind that reliable air coverage of the entire Los Angeles basin can be accomplished" within a 40-mile radius of Mount Wilson.[123]

After surviving a superior who nearly killed the idea, Silva suddenly had the opposite problem: a boss who wanted it *now*. Schulke asked almost every day when the flying TV station would be completed. With a gut feeling that his chief engineer would make television history, the general manager also assigned photographer Dave Kovar to the project. "I'd like him to track you through this whole thing and document what's going on," Schulke told Silva.

Deciding that KTLA couldn't afford the larger choppers on the market, Silva rented a Bell G-2 helicopter. But he couldn't maintain secrecy if he worked in full view of other pilots at the Van Nuys airport. Dick Hart, the owner of National Helicopters, offered the big backyard of his Studio City home as a work area. Fortunately, no one bothered to ask why they were driving across the San Fernando Valley in broad daylight with a helicopter hitched to a tow truck.

After experimenting with every known type of antenna, Silva visited the General Electric plant in upstate New York in May 1958. He liked the helical model GE had developed for UHF television stations. Its 360-degree signal would provide freedom of movement for the chopper. But it was still too big. GE engineers modified the design and shipped the parts to the Paramount machine shop for assembly. "When we got the equipment I had to keep bringing the weight down," Silva said. "I had this great big scale at Paramount Pictures, and I'd put the equipment on and keep taking stuff off. People said, 'He's a nice guy but he's crazy.'"

The end-fed helical antenna spiraled along the length of a five-foot rod, three inches in diameter. The antenna was held in place by knobby ceramic insulators and spacers. The contraption had to be fully extended beneath the helicopter in order to transmit a TV signal, but it hung down below the skids of the chopper and made it impossible to land.

"If you landed, you would pole vault," Silva explained. "You had to have a way to get the antenna out of the way of the skids. The only way I could figure was to put it on a pivot. Paramount came up with this mechanical device where I would have a handle to pull it up in three pieces. I could reach down, grab the handle, click one third up, then reach down, click two thirds up, and then on the last stage, click and we could land."

Even as the weight of the gear was reduced, there was no way to cram everything into the bell of a two-person helicopter. Silva designed long black boxes for the skids, which provided storage space and kept the chopper balanced. The project still didn't have a name. Rejecting "eye in the sky" as too cute, the engineers agreed on a suggestion from reserve pilot Larry Scheer: the KTLA Telecopter. They painted it in large letters on the skid boxes, adding a big disk with the channel number 5 on the tail.

Dave Kovar kept snapping pictures. The photographs show the engineers installing the transmitter in the skid boxes and wiring the equipment against the leafy backdrop of a suburban backyard. They don't show the men sweating in the intense midsummer heat of the San Fernando Valley as they battled to reduce the noise and vibration. Keeping the picture steady and focused required another secret deal with a company that made gyroscopic platforms for movie cameras aboard World War II aircraft. "Jim (Schulke) put a lot of pressure on me to do it and document it," Silva would later recall. "But it was still cloak and dagger, the whole thing, like a dime novel."

Surprisingly, it took only three days to put all the pieces together. A special rotating dish was installed on Mount Wilson to receive the signal. Once again, the "secret" project was towed through the streets, back to the Van Nuys airport.

***

On July 3, 1958 the Telecopter was ready for its first test flight. The TV gear now weighed a mere 368 pounds. Pilot Bob Gilbreath carefully gained altitude. Silva transmitted a signal. Mount Wilson received nothing.

"I was very sad that day because we didn't get (a picture) out," said Silva. "I was having trouble fighting the heat, vibration and noise. I had to replace the tubes in the transmitter and put in some absorbers to get rid of the vibration." The only way to adjust the gear was to hover at 5,000 feet with Silva balancing on the skids. "I had to climb out on those things when we were testing—with no seatbelt—and try to fix some of the equipment," Silva recalled. "The project is very dear to me because I put my job on the line and I put my life on the line. There was a lot to lose and a lot to gain."

Everything was ready for another test on July 4. This time Mount Wilson received the picture. Silva and his fellow conspirators were elated. They waited for KTLA's general manager to request the daily progress report.

"How are you doing, John?" Schulke asked.

"Ask me again."

"When will you be ready?"

"Now!" Silva exclaimed.

***

On July 24, 1958, KTLA summoned print reporters and public officials to the Los Angeles Police Academy. According to Silva, "We didn't tell them what we were doing, just that we had a surprise, a secret to show them."

The newspaper scribes gathered around a closed-circuit TV monitor. KTLA news director Gil Martyn announced that they were about to see the world's first TV news helicopter. The picture would tell the story. Bob Gilbreath piloted the chopper in a wide circle around the Los Angeles civic center. Silva aimed his camera and focused on the building. To keep the picture from jiggling, he was careful not to zoom in too close. Years later, the inventor would brush a tear from his eye as he recalled the moment. "The first thing we showed was city hall and they all broke out in applause. We came in to land right there just a few feet away from everybody and we were still showing pictures as we came down." The reporters could see themselves in the live picture as the chopper landed. Silva had to put aside his emotions and remember to retract the antenna. One... two... three clicks. They were safely on the ground with the astonished journalists and dignitaries.

Daily Variety hailed Paramount's $82,000 investment as the "newest razzle-dazzle innovation in television." [124] TV-Radio Life declared, "Local TV and radio stations continue to lead the nation in news coverage. First there were the radio freeway reports, then KABC's radio helicopter and this week KTLA launches its daily Telecopter newscasts..."[125] The Los Angeles Times added, "The advantage of being able to reach the scene of a disaster—a major freeway pile-up, a train wreck—through the air above the tangle of Los Angeles traffic is obvious."[126]

When print journalists questioned how viewers would know when to turn on KTLA for breaking news, Silva had an answer. The Telecopter was equipped with a flashing light on the bottom that could be seen 30 miles away in daylight. The idea was for Southern Californians to see the bright light in the sky as a signal to run indoors and turn on the television. Reporters and civic officials at the helicopter unveiling were asked to embargo the story for one more week, so the publicity would coincide with the first live broadcast from the

Telecopter. No one talked about the possibility that the invention might crash on live TV.

The July 29 broadcast called for the Telecopter to land in the KTLA parking lot where a "mock news story" had been set up. "We were going to come in and land and take pictures on the way down and show it to the public," Silva said. "I showed pictures of the freeway and we came in." Stan Chambers was on the ground to interview the Telecopter crew. They swooped down for a landing.

A few feet above the concrete, Silva heard a sickening thud and felt the chopper lurch and wobble to the side. "I'd been doing this for a week. When we were doing the program I was so busy trying to take pictures that I forgot to do that little thing of pulling the antenna up in three stages before we landed. And we came down and we hit that thing..." He braced himself for what he thought was an inevitable crash.

Bob Gilbreath called on every moment of his piloting experience and steadied the airship. "We were ready to pole vault right over the antenna," Silva recalled. "A less experienced pilot would have crashed right there. But he pulled it up and I pulled the lever up and we landed. Everybody on the ground went, 'Oh God, it's broken.'" Worse, the near-crash had been shown on live television. Disappointed, General Manager Schulke told Silva they could try again the next day.

"I think it will work, Jim," Silva said, not wanting to abort the Telecopter's broadcast debut. Gilbreath started the engine and lifted off from the parking lot. Silva lowered the antenna and aimed the camera out the window for another bird's eye

Inventing TV News

view of traffic on the Hollywood Freeway. "Mount Wilson reported a signal, so we carried right on and we did our program. Didn't even have to repair it at all. That was the story of the first day. And every day after that we were on the air."

Silva told TV-Radio Life magazine that KTLA had "exclusive rights to the helical antenna and to the name 'Telecopter.'" He was confident other Los Angeles stations could not copy his invention simply by putting a transmitter in a whirlybird. "They would have to follow a rigid flight pattern," he said. "It would be a very expensive operation, and besides, nobody likes to be second." KTLA offered to help stations outside Los Angeles build their own TV helicopters, but there were no takers.[127]

As he had promised the technical unions, Silva handed the camera and engineering duties over to Harold Morby. Test pilot Gilbreath had to return to his regular job with National Helicopter, so Larry Scheer assumed the duties of flying commentator. Time Magazine reported, "The two-man crew was picked for their light weight and warned to stay thin."[128] Morby continued to work with Silva on constant improvements to the technical gear. A highly skilled pilot with some prior experience as a journalist, Scheer defined the role of the airborne reporter as he went along.

"Originally, they would give us a news assignment, and they would dictate a story to me over the telephone and I would write it down longhand," Scheer told a KTLA publicist. "We would go out and I would try to read (the script). It got to where this wasn't satisfactory because I couldn't fly the machine, read, watch the monitor and keep track of what was

115

going on (on the ground). There was no way in the world this could be done. So, I decided to take notes, and that didn't work either. I have a pretty good memory, so I said why don't you give me the basic facts over the radio and give me the opportunity to put a story together and see how it comes out. Even on breaking stories, on a fire they would try to dictate a story to me on the telephone. (It) was basic, but not what I saw. It was factual as to the amount of (fire-fighting) equipment, the acreage and when it started, but it still included nothing of what I could see when I was there. When I started reporting what I could see and interlocking that with the facts that they had given me, then we started to get a decent news story. It also gave me more freedom of action to fly because I didn't have to look at a piece of paper.[129]

"I have no copy written for me. I write my own as I go," Scheer explained. "Everything is extemporaneous, it's ad lib or it's as I see it. There's no time to read copy… This also makes me unique in that I am probably the only totally unedited reporter in the whole world…" The pilot-reporter also had sole responsibility for deciding when weather conditions made it too dangerous to fly. He also had the option of feeding only video, if he determined that doing an audio commentary would distract him from the task of avoiding obstacles and other aircraft. In that case, the aerial shots would be narrated by an anchorman in the studio.

Telecopter reports were recorded on videotape for possible later airings but the technology for videotape editing had not yet been developed. Scheer had to edit his reports in the sky. "I have trained myself… to do my stories in the period of time that the station says they want it done in. They say, quite frequently, we want 45 seconds filled, and I don't care how important the story is. If they want 45 seconds, I cut it to 45 seconds. If they tell me they want two minutes, I don't care how insignificant the story is, I try to make it last for two minutes. I do perform an editing job... either stretching it or cutting it..."

Daily use of the helicopter in newscasts also played a part in branding KTLA as "The Telecopter Station," distinguishing Channel 5 from other independents and network-owned competitors. According to an unpublished station history, "The telecopter is not held in reserve for major disasters, it is used on a daily basis to report the everyday occurrences in this giant city."[130]

Stan Chambers recalled one assignment that didn't turn out as planned: "The circus was coming to town and as part of the hoopla promoting the arrival, ten elephants were to parade down Hollywood Boulevard. The chopper was sent up to show pictures of the giant beasts marching trunk and tail through the city." When the crew was unable to locate the parading pachyderms, "News Director Gil Martyn muttered, 'How do you like that, 10 elephants in the middle of Hollywood Boulevard and you can't even find one of them with the Telecopter.'"[131]

The KTLA helicopter's early assignments in the late 1950s foreshadowed Southern California's obsession with televised freeway chases in the 1990s. The chopper team covered police pursuits, a plane landing on a freeway, even a suicide jumper being talked off the ledge of a high-rise building. If nothing was happening, aerial views of Southern California provided endless scenic backdrops for weather and traffic reports. The view from above wasn't always grim. A hapless lady golfer was caught in the act of picking her ball out of the rough and tossing it toward the green, much to the merriment of the chopper crew. The eye in the sky also raised questions of privacy as it proved useful for snooping in celebrities' backyards.

"KTLA's Telecopter scored another news scoop last evening when the flying television station hovered over Marlon Brando's Mulholland Drive mansion just 20 minutes after Mrs. Brando had discovered her Japanese maid drowned in the swimming pool," a Paramount TV Productions release reported on September 12, 1958. "Fifteen minutes after receiving the tip, KTLA's tele-copter was on the air with picture and sound from the scene of the tragedy. Viewers saw the body being removed from the pool and transferred to the ambulance."[132] However, the story was regarded as having little news value at the time; Brando said the 31-year-old housekeeper was learning to swim and apparently strayed into the deep end of the pool. The Los Angeles Times covered the drowning in three paragraphs, devoting more of its front page to the fractured marriage of actress Debbie Reynolds and crooner Eddie Fisher, who was reported to be having an affair with movie star Elizabeth Taylor.[133]

Like Silva, Scheer believed that "the objective of the Telecopter is primarily to get to the scene of a major news story first and get it on the air first—factually—that's the whole objective of the Telecopter. Anything else we do is coincidental." But law enforcement was quick to see the advantage of instant live pictures overlooking crime and fire scenes. A KTLA historian noted, "The Telecopter served a two-fold purpose. It provided the viewer with instant news and it served as a monitor for the law enforcement and fire departments."[134] This cooperative relationship between police and airborne broadcasters raised no journalistic eyebrows in the late 1950s. In fact, the station boasted of assisting law enforcement as a public service.

Paramount's technology investment also improved KTLA's bottom line. Broadcasting magazine declared, "There's money in newscasting... but KTLA shows you can't scrimp to make it." Station employees posed for a 1959 photograph, showing "equipment worth $443,500 and a $175,000 a year news staff." The KTLA crew proudly showed off their TV toys: the Telecopter and the Telemobile, a ground unit in a GMC truck that could broadcast pictures while in motion. The station also had two remote trucks and a four-camera production van in addition to news film equipment.

Broadcasting said the station counted $728,000 in annual revenue from its two half-hour, Monday-through-Friday evening newscasts, with additional income from daytime and weekend news shows. Citing the station's Golden Mike awards for news coverage, Schulke said KTLA's news operation was creating "the kind of reputation that pays off on both rating sheets and financial statements."[135] This was a revolutionary idea at a time when most stations lost money on news to meet the FCC's public service requirements.

But profits took a back seat to public service when KTLA pre-empted all programming and commercials to cover a crisis. Schulke stated, "It is incumbent upon a television station not to merely offer entertainment but... to serve the people of the community continuously and with integrity... It is our hope that the Telecopter will not only provide on the spot news coverage but serve the community in times of emergency."[136]

It didn't take long for that bold prediction to be tested.

***

The golden hills of Southern California provide an idyllic setting for luxury homes. But residents can pay a steep price for the view. In the fall, hot Santa Ana winds ruffle the tinder-dry brush. A tiny spark can quickly grow into an inferno.

Herb Green was first to notice the smoke. It was 8:16 a.m. on Monday, November 6, 1961. Green was flying a helicopter during the morning rush hour to provide traffic reports for radio station KMPC. The small fire in Stone Canyon didn't look like much of a story at first. No one even bothered to roll videotape when Morby and Scheer aired their first live report from the KTLA Telecopter at 8:30. They aired three more updates over the next hour. But when newsman Clete Roberts arrived on the scene with the Telemobile at 9:30, he took one look at the wind-whipped flames and knew it would be a disaster.[137]

"This mesquite, this oily brush on the hillside is literally packed with oil and it explodes. Flames can move at the speed of the wind. We get 50, 60 mile-per-hour winds," Roberts recalled.[138] "I was one of the first reporters to go into the field.

We could see the pall of smoke. We could smell it. It was low to the ground. It was as though a dark cloud had passed over the Southland sky. There were a lot of embers dropping out of the sky and into the street. We began to see people coming out of the hills."

The ruggedly handsome Roberts had left CBS to join KTLA in 1959. His swashbuckling image as a foreign correspondent was reinforced each night as the opening of the newscast showed the anchorman in a trench coat, passport in hand, stepping off an airplane. According to Stan Chambers, Roberts modernized KTLA news by stressing international stories and drawing on his worldwide network of newsreel cameramen. "It was during Clete's years at KTLA that the picture took over," Chambers said. "When a breaking story developed, Clete Roberts knew how to get the most from the live camera. He had the ability to make the viewers feel that they were intimately involved in the story."[139]

***

Observing the fire from the Telecopter, Scheer warned, "Many expensive homes are in immediate danger... (some of them) worth as much as $60,000 dollars!" The houses in the path of the flames belonged to some of Hollywood's elite; they were running for their lives in their Cadillacs, Lincolns and Thunderbirds.

Clete Roberts reported that there was only one road out of the canyon. He questioned the fleeing residents. "One guy said, 'I've got this bottle of scotch,' and his wife said, 'I've got the pictures of the kids and the marriage license.'"

Back in the newsroom, John Hilliard, an unglamorous newsman with thick glasses, was standing in front of a map with a felt-tip marker to pinpoint the location of the flames. "Take a look at the map for just an instant," Roberts said. "Then come back here because these houses are going to go."

From his rented house at 901 Bundy Drive in Brentwood, then-former Vice President Richard Nixon watched a neighboring home go up in flames. Nixon and his chief researcher scrambled up onto the roof to water down the shingles, but turned on the garden hose and discovered there was no water pressure. They grabbed their manuscript for the book "Six Crises" and obeyed police orders to leave the area. Fortunately, Mrs. Nixon was running an errand that morning and their daughters were in school. Left behind was the family dog, Checkers, who would be rescued later by newsmen.[140]

Filming of motion pictures and television shows came to a halt as celebrities rushed from the studios to their burning homes. Kim Novak and James Garner left the set of "Boys Night Out" to water down their roofs. Robert Stack, star of the TV series "The Untouchables," raced home to save his gun collection. Flames surrounded the home of producer Irwin Allen, who would become famous for epic disaster movies like "The Towering Inferno."[141] One hundred studio employees tried to save the home of Fred MacMurray, star of the TV sitcom "My Three Sons." MacMurray and his wife drove away with two carloads of valuables. But when they tried to return for their twin five-year-old daughters, the road was blocked. They ran the five blocks in time to rescue the children, their babysitter and the family dog before the house caught fire.[142]

The Red Cross opened shelters for fire victims but most of the upscale evacuees fled to luxury hotels. They had become the accidental stars of a real-life TV drama.

KTLA would pre-empt regular programming and all commercials for the next 11 hours and continue regular cut-ins throughout the night. The Los Angeles Examiner called it "a decision that was both costly and admirable... a display of responsibility and news judgment all too rare in TV." Newspaper columnist Charles Denton noted:

"The other stations tossed in the towel without a struggle, ignoring the holocaust except for occasional bulletins and trying to amuse the burned-out evacuees with what passes for daytime programming... I found it bitterly laughable to see Johnny Carson and 'Amos and Andy' gagging it up while a small section of the world was crinkling in flames like cellophane in an ashtray. Meanwhile, back at KTLA, television journalism was taking a long stride toward maturity under the sure though wearying guidance of Roberts and his colleagues."[143]

***

KTTV was the only other station to pre-empt programming for fire coverage with Bill Welsh at fire headquarters, three remote units and four KTTV film crews. "Welsh tried to make a fight of it in his own pathetic style, bustling about interviewing everyone available. But he came too late with far too little," columnist Denton wrote. Independent KHJ and CBS-owned KNXT offered their crews to KTLA's marathon effort but didn't interrupt their own programming. The film on the network evening newscasts couldn't compete with the live drama on KTLA.

Roberts interviewed a frantic woman motorist. She begged to be allowed into the fire area to rescue a girlfriend's father-in-law. Authorities assured her that the man already had been accounted for. Seeing that she was too distraught to drive, rescuers helped the woman move her car.

Later, Roberts was on the ground with the fire consuming homes on both sides of the street. "Are they telling us to get out of here?" he asked. He narrated the relentless march of flames toward a hillside house as the homeowner apparently gave up and walked away. "That house is going to go," Roberts said.

Mobile ground cameras showed residents leaving their homes, carrying everything from chairs to shotguns. KTLA General Manager Stretch Adler had joined Roberts and cameraman Dick Watson in the field. Viewers watched the TV men helping to fight the flames as they covered the story. "An effort is going to be made to save this house," Roberts said. "All right, we're going to put this fire out."

But it was a losing battle. Roberts managed to save the family's silver tea set and deposited it on the lawn. A TV cable became stuck under a fire hose. A fire department pumper truck didn't have enough water pressure to make a difference. While Roberts was trying to save the homes of strangers, aerial tankers were dousing his own house in Benedict Canyon, which was also in the path of the flames. Roberts' wife was among the evacuees.

Inventing TV News

While some found Roberts' reporting "theatrical," Stan Chambers called it "one of the most powerful and dramatic telecasts ever seen. The marathon coverage of the fire's unbelievable devastation was the telecast that first brought the visual powers of the KTLA Telecopter to millions of viewers in Los Angeles and across the country." Spectacular pictures showed the air war against the fire as "borate bombers" dropped their loads. As night fell, the view from the sky showed the tiny silhouettes of firefighters locked in futile combat with giant flames. People clogged the KTLA switchboard to voice their appreciation for the coverage. Newsmen made an on-air appeal to viewers to stop calling and free the lines for emergency use.

The city editor of the Los Angeles Mirror wrote to Clete Roberts, "KTLA's fire coverage was the best thing of its kind I've ever seen on television. First time I've witnessed the ultimate exploitation of the medium." Another note from the editor of the Pasadena Independent Star-News added, "Your coverage of the fire has been magnificent. You beat KTTV easily. The rest were pathetic. You have shown what TV can do on a major story and should be extremely proud."[144]

The next day brought more coverage and new danger. A KTLA remote truck became trapped in the flames. Engineer Ray Phillips sought refuge in a garage, which also caught fire. According to a Paramount press release:

"Phillips found himself in the center of a circle of flame with his truck afire. At great personal risk he managed to douse the flames with a fire extinguisher. However, when he went to start the truck, the motor would not turn over because the fire had consumed all the oxygen. For 30 minutes he stood by, trapped in an inferno engulfing, in huge swallows, homes, cars and garages on all sides. It was only an Act of God that made the fire, in its furious charge over the terrain, by-pass Ray and the truck. Soon after, he and another member of the crew managed to start the truck and off they went to a new location in the performance of their duties."[145]

***

By Tuesday night the Bel Air fire had consumed 14,000 acres, destroyed 456 houses, caused more than $20 million in property damage and driven 3,000 people from their homes. Incredibly, no one was killed. But stars like Zsa Zsa Gabor were among those who lost their homes in what, at the time, was the worst fire in Los Angeles history. Actor Burt Lancaster and his 10-year-old daughter returned to their $500,000 Bel Air mansion to find only a chimney and charred trees. "The firemen did as good a job as possible under the circumstances," said Lancaster, thankful that his entire family was safe. "To stop it they'd've needed half a million men."[146]

KTLA had been on the air for more than 18 hours. The decision to pre-empt programming and commercials cost the station an estimated $39,000 in lost revenue but drew a huge audience. [147] Twenty-three percent of nearly 2.2 million television households in Los Angeles were tuned to fire coverage on KTLA, with almost half that number watching KTTV.[148] The networks asked, "Who stole the audience?" as the estimated audience for a Monday night variety special was cut in half by the crowd of Angelenos watching KTLA.[149]

TV Guide raved, "For millions of television watchers it was a tremendous and unscheduled spectacular in which they shared the front-rank struggles of the firefighters and the heartbreaking losses of those who watched their homes go up in flames."

Mayor Sam Yorty commended the station for "public service of the highest order and highest praise." Fire officials credited the station for saving lives by keeping sightseers away from the disaster zone, "satisfying the appetites of the curious and relieving the fears of the concerned."

"It takes a catastrophe to bring out the best in local television," wrote the Examiner's Denton. "But the fact remains that the local lads attain their highest level of service when nature affords them a multi-million-dollar production free of charge."[150]

"This is where we're most potent, on fires and disasters," said Telecopter pilot-reporter Larry Scheer. "We are primarily a disaster machine, because we can always get to the scene of a disaster and get there first." The only reporter in the skies over Los Angeles was aware that he still had to beat the competition on the ground. "We do it live while others do it late. A lot of times we do the same stories as our film crews and to me it's better when we've done it than when our film crew has done it. Ours shows up on the five o'clock news because the film hasn't been processed yet."

But Scheer also realized the limitations of the bird's eye view: "The film crew is able to get close up on the target, they are able to get interviews... shoot different angles... give you a broader, more customized view of what the news story is. We only have one angle, looking down. We don't have the opportunity to talk to the individuals involved, so our story is a different story from the one on the ground... I think it's more exciting coming from the air."

Scheer also realized that the helicopter could merely witness the destruction, not probe for the cause of the disaster. "I cannot take someone over the coals, I cannot puncture someone's story. I merely have to say that Joe Blow stated such and such or did such and such... If I were interviewing the man while he said it as a fact I might be able to interrogate or bust his statement to where what he said... was not true. So the ground crew has different advantages to me."

***

On November 22, 1963, Americans gathered around their TV sets to watch the live coverage of the assassination of President John F. Kennedy. "In a completely unprecedented act of cooperation, the television networks permitted (independent stations like KTLA) to join them in their coverage of the assassination," Stan Chambers recalled.[151] "It was one of those rare moments... when every station in the country was broadcasting the same story." It would go down in history as the day network television news came of age. Only a few weeks later, the nation would be watching another disaster unfold in Los Angeles, with KTLA's Telecopter and Telemobile once again on center stage.

***

Inventing TV News

It began as a tiny crack, barely visible in the massive earthen dam in the Baldwin Hills, south of downtown Los Angeles. Hardly worth a mention in the daily newspapers, which were pre-occupied with the kidnapping of 19-year-old Frank Sinatra Jr. The local TV newsmen were still laughing at how Ol' Blue Eyes himself had tricked the press by serving them a candlelight dinner outside his Bel Air home. The elder Sinatra had paid a $240,000 ransom for the release of his son. While the newsmen were feasting on an elegant spread, catered by Chasen's restaurant in Beverly Hills, young Frank was spirited past the press corps in the trunk of a friend's car. Nice meal, but no interviews. Across town that same evening, unnoticed by the media, the tiny crack in the dam was about to widen into a life-threatening emergency.

On Saturday, December 14, 1963, a caretaker for the Los Angeles Department of Water and Power was inspecting the Baldwin Hills dam. He heard a rushing sound, an unusually heavy flow of water in the runoff lines. Further inspection revealed a deep, jagged line in the dirt, no wider than a pencil. A dozen men began trying to plug the crack, which zigzagged from the top to the bottom of the 66-foot wall. Other workers opened the discharge valves to release water, hoping to ease the pressure on the dam.

Police went door to door, advising residents below the wall that they might have to leave their homes. Many ignored the warning at first. They couldn't see that the danger was real. By 2 p.m. the crack was widening, parting like a theater curtain. With KTLA's Telecopter overhead and the Telemobile parked on a bluff with a panoramic view above the water line of the crumbling dam, the disaster seemed to unfold right on cue.[152]

***

Pilot/reporter Don Sides narrated from the helicopter: "The water broke over the top, filled up the lower catch reservoir, and right there are two homes that are completely engulfed. One of the homes, the walls are washing away at this very time right now."

The videotape of the live report, included in a same-day special report examining the possible causes of the dam collapse, shows that Sides' narration often did not match the pictures he was transmitting. Sides was the newly hired replacement for pilot/reporter Larry Scheer. It was the new guy's first time covering a major story from the chopper. Operating the camera was engineer Lou Wolfe, filling in on Harold Morby's day off.

"We have been on only five minutes and there's probably 50 homes already destroyed completely," Sides went on. "We will swing up and show you the hole and the tremendous suction of water that's being pulled out of the Baldwin Hills reservoir. The whole dam is about ready to go at this point."

The Telecopter camera zoomed in on the failing dam, just as a giant piece of concrete broke off. KTLA then cut to a live shot from the Telemobile, showing chunks of concrete bouncing on waves of floodwater. At this point, the audio from the live signal was lost. Anchorman Joe Benti pointed out, "The pictures at this point did not need description. As if a

Inventing TV News

bomb had been set off, the earthen dam ruptured in a torrent of cascading water."

Sides reported that police on the ground were "trying to activate people so there won't be any more loss of life. The homes are going out so fast you can hardly count... just moved off the foundations, off the lots and down the street." He also noticed that the skies were filling with military and rescue helicopters, "not there to take pictures but to rescue those trapped in their homes and apartment buildings."

"People inside were advised to go to the roofs," anchorman Benti stated. Hundreds were airlifted out as the Telecopter and Telemobile continued to transmit pictures of the destruction. In the aftermath, as the devastated neighborhood was left without power, ground crews documented the small details.

"In one house a cat sat shivering. In another only the battered kitchen stove was left behind," said KTLA reporter Tom Snyder, estimating that 64 homes were washed out, 96 apartments damaged and 117 other structures destroyed. Three people were killed.

KTLA reported that authorities later examined the footage from the Telecopter as they tried to pinpoint the cause of the disaster. News footage showed bespectacled engineers in a conference room at the TV station, watching the videotape. The helicopter also flew over other Southern California dams located near residential areas as journalists raised the question, "Can it happen again?"

Los Angeles Mayor Sam Yorty commended KTLA's coverage. Channel 5 "did a wonderful job... your continuous show. We were watching down at command headquarters. It was useful in another way that maybe you didn't think of... it kept a lot of sightseers out of here that could have caused us a lot of trouble. There were enough as it was." Independent KTLA shared its coverage with all three networks. The Los Angeles Times wrote: "KTLA has consistently proved that it is without equal in live coverage of local news."

***

The Telecopter crew would face a more complex challenge during the 1965 Watts riots, spending more than 30 hours in the air for continuous coverage that was closely monitored by the police, and by demonstrators on the ground as well.

"To avoid sniper firing, the Telecopter flew its missions over the fire-ravaged Watts without lights... (and) in the flight path of the Los Angeles International Airport," Rotor and Wing magazine reported. [153] Reports from the chopper pointed out specifically where police should go to arrest suspected arsonists and looters. Aside from the ethical conflict of journalists serving as an airborne lookout for the police, the aerial pictures skimmed over the most dramatic scenes without shedding light on what prompted the violence. KTLA's coverage, however, was honored with the prestigious George Foster Peabody Award.

Despite the obvious competitive advantage, throughout the 1960s no other U.S. television station matched KTLA's financial and technical commitment to live reporting from the air. By 1970, KTLA was flying a Bell 206 Jet Ranger, customized at the Bell helicopter plant in Fort Worth, Texas and described by an aviation trade magazine as "the world's first and only flying television station that can transmit on-the-spot, live, black and white or color pictures." KTLA management took pride in sharing those pictures with American and worldwide TV networks, and even competing local stations, when big stories broke, such as the 1971 Sylmar earthquake. It was the kind of public service that FCC regulators wanted to see when renewing the station's license. Why didn't other stations follow suit?

By early 1970s standards, helicopter coverage came with a hefty price tag. The Jet Ranger was Silva's fourth model, broadcasting in color with $330,000 worth of equipment and an operating budget of $82,000 per year. Invaluable during live coverage of a disaster, it was harder to justify the cost on slow news days when the chopper was assigned to "routine feature stories and traffic jams on the freeways plus girl-watching along the beaches." KTLA was now owned by Golden West, where former cowboy film star Gene Autry served as chairman of the board. The news department could no longer count on the deep pockets of Paramount Pictures.

***

As stations replaced their film cameras with electronic news gathering on videotape, the eye in the sky no longer provided the most dramatic live pictures. This was demonstrated on May 17, 1974, when LAPD tactical units confronted members of the Symbionese Liberation Army in a fiery gunbattle that killed most of the suspects in the kidnapping of newspaper heiress Patty Hearst.

"We tried to use the Telecopter at the SLA shootout," said KTLA General manager John H. Reynolds, "and we couldn't get in close enough to do anything... our best stuff was taken on the ground."

In 1974, KTLA sold the nation's only flying TV studio to rival KNBC for a reported $350,000.[154] Although Reynolds insisted that the sale did not result from any financial considerations, Silva had a different view. "It was economics. We had a general manager that decided we were going to cut costs by getting out of the real big stuff. One day he called me into his office and said, 'I'm sorry, we're going to get rid of the 'copter.' He actually made the deal with Channel 4 and then told me, even though it was my baby. I had to go over to KNBC and show everyone how to use it. They were awkward in the beginning but they got better."

As part of the deal, reporter-pilot Larry Scheer also switched stations. But KNBC wanted a bigger-name to fly its news chopper. The station hired Francis Gary Powers, the Cold War U-2 pilot who was shot down over the Soviet Union in 1960 and convicted of espionage, spending two years in a Soviet prison. He began flying for KNBC in 1976, after a brief training course at Bell's helicopter school in Texas. Channel 4 ran ads and even distributed playing cards with his picture, leading some to criticize NBC management for overlooking his relative inexperience as a helicopter pilot. The Los Angeles Times reported that Powers developed a reputation for risk-taking. "It was an in-house joke that he thought he was still in the U-2 and flying high-altitude reconnaissance," the newspaper said.[155] Powers was killed on Aug, 1, 1977 when he ran out of fuel and crashed the nation's only live TV helicopter while returning from a news assignment.[156]

***

By the time it sold the Telecopter in 1974, KTLA boasted that its aerial coverage had won 40 local and national awards. Yet, the station abandoned the high ground to a competitor, citing cost issues just as television newscasts were beginning to emerge as a profit center for other local stations in the 1970s. "KTLA could no longer justify the expense in the face of heavy news competition from the three network-owned TV stations in Los Angeles," Broadcasting magazine reported.[157]

Ironically, it was KTLA's market niche as an independent station that made possible the development of the Telecopter. Unconstrained by a mandate to carry network programming, KTLA could pre-empt its local programs at any time for live, breaking news. Helicopter reporting would not take center stage again until the mid 1990s, when local TV stations sought ways to compete with cable news offerings and the internet. Aerial video from competing TV helicopters drew huge ratings during live coverage of high-speed police pursuits on the L.A. freeways.

Six decades after the debut of the KTLA Telecopter, it is difficult to imagine a major competitive local TV news operation without access to a helicopter for live traffic and breaking news reports. However, the same limitations are also present: aerial footage offering only a superficial view of what's happening on the ground and emphasizing action and drama over context and substance.

Chapter Five:
The Real Ted Baxter

The Rise of the Celebrity Anchorman

> *I became... a different breed of celebrity, one who was a television news reporter... I got the credit although it was television that really made everything happen.*[158]
>
> —Stan Chambers, KTLA newsman

The young announcer did not yet have a national reputation. He had not been seen on television outside of Washington D.C., where he read the 11 p.m. news on the local CBS-affiliated station, WTOP. However, "he was warm and friendly and had the unusual knack of being able to communicate with his viewers."[159] The general manager of WTOP didn't want to release the newsman from his local TV contract when the network came calling. But CBS executives were convinced that this former war correspondent for United Press International had a "strong enough personality to earn the full respect of the viewers." His name was Walter Cronkite.

'Anchor' is a uniquely American way of referring to the job of reading or presenting the news on television. The term was coined in 1952 by CBS executive Sig Mickelson to define

Cronkite's role in network coverage of the national political conventions. As he ascended to the anchor chair of the CBS Evening News, "the most trusted man in America," [160] personified the image of the credible network newsman, knowledgeable about the issues and dedicated to the highest ideals of broadcast journalism. Despite the success of the Huntley-Brinkley dual anchor format on NBC and a later, three-anchor format on ABC, the image of the white, male news presenter—the TV descendants of Walter Cronkite—presided over the network evening newscasts for five decades. In the aftermath of the traumatic September 11, 2001 attacks, Americans turned to trusted anchors Dan Rather of CBS, Tom Brokaw of NBC and Peter Jennings of ABC.

On the local level, popular culture developed a very different image of the anchorman: Ted Baxter, the pompous newsreader for fictional Minneapolis station WJM on a popular TV sitcom, the Mary Tyler Moore Show. Ask almost anyone who was working anywhere in local television news in the 1960s and 70s and they'll swear that *their* station's anchorman was the *real* Ted Baxter. The local anchor was also lampooned by Chevy Chase and others on the "Weekend Update" segment of NBC's "Saturday Night Live," by the animated character of Kent Brockman on the Fox TV series "The Simpsons," and by Will Ferrell in the 2004 film comedy "Anchorman: The Legend of Ron Burgundy." Why did the network television news anchor evolve as an icon of respectability, while the local broadcast news anchor came to be regarded as a blow-dried buffoon?

Part of the answer lies in Los Angeles, where the creators of the Mary Tyler Moore Show drew their inspiration from two familiar personalities on the local news, George Putnam and Jerry Dunphy. But the image of the local anchorman evolved with television news from its beginning. As film director and TV historian Michael Ritchie has observed, "The extraordinary fact of the slow development of television news was that it took a long time for everybody—programmers, advertisers and viewers,—to realize the full impact of a statement of 'fact' by a face on a television tube."[161]

***

In the early days it was not obvious that television news anchors would someday command million-dollar salaries at the networks and in the largest TV markets. As late as 1949, some stations did without on-camera talent, simply broadcasting the text from a wire service news ticker. An off-camera announcer might narrate film footage or read wire copy while a 'news' slide filled the screen. "There was no attempt to gather news or rewrite the wire copy," recalled George Eisenhauer, who read the news on DuMont's WDTV in Pittsburgh. "Everything was ripped and read... but the weather, and you looked out the window to get that."[162]

Most early TV programmers envisioned some type of on-camera news presenter but the exact nature of the role was open to debate. "Should the central figure be a ringmaster to drive the program forward or a guide and interpreter...a father figure, a show business personality, a star, a widely known reporter, or a competent news reader? No one was quite sure,"

Mickelson recalled. "The staff tried an elderly man with a beard, an aggressive young sportswriter from a New York daily newspaper, and finally a number of staff announcers who were professional performers. It was quickly determined that they still did not have an answer."[163]

The most successful announcers were on the radio and their skills didn't always translate well to the new medium. One radio broadcaster appeared on a local television news program in 1939 and was panned in the New York Times: "...a new, more informal and natural style must be developed for such telecasts. Spectators agreed that reading, with his head bobbing up and down and occasionally looking up from the paper did not fit the intimate medium of broadcasting."[164]

The popularity of radio in the 1930s and 1940s made it hard to imagine that a television news program would ever draw a bigger audience than one of Edward R. Murrow's dramatic wartime reports. Although most local radio news announcers followed the 'rip and read' format, the emergence of an elite corps of 'Murrow's boys' at CBS added a veneer of respectability to radio. Top radio journalists sometimes doubled as voiceover announcers for motion picture newsreels, but they shunned television. CBS officials took this into account when choosing Douglas Edwards as their first TV news announcer. "It was clearly necessary to select one of the lesser lights who would feel he had nothing to lose by being identified with an environment that was considered too frivolous by the elite of the staff," Mickelson explained.[165] For new talent, the networks looked to the few local TV stations in operation.

John Cameron Swayze debuted as "the first regular newscaster on TV" in 1937. Experimental W9XAC in Kansas City aired a ten-minute news program three days a week. The announcer was "a soggy mess" when the show was over. The lights were so hot that the newsman painted his eyebrows black to keep them from being singed off. "The work was relatively simple," said Swayze. "I just read the news right out of the paper."[166] Swayze later moved to New York and became nationally famous as the anchor of NBC's Camel News Caravan. Swayze was a non-smoker but many viewers thought he smoked because the show concluded with the image of a burning cigarette in an ashtray.

Swayze memorized all of his copy and used catch phrases such as "hop-scotching the world for headlines." Meticulous about his appearance, he wore a toupee, changed his tie every day and wore a flower in his lapel, much to the annoyance of colleagues who considered him "more a performer than a newsman."[167]

A young Chicago radio announcer named Hugh Downs was asked to read the news on WBKB-TV in September 1943, when there were only 400 receivers in town. He was handed a thick sheaf of copy and told he had 15 minutes to deliver it on the air. He could see his breath as he sat down at a plain desk and chair on the freezing set. "Boy, this place is chilly," said Downs. The female producer came in wearing an Eskimo-style fur parka and told him it was okay for him to take his jacket off and do the broadcast in shirtsleeves. Downs didn't get it. His lips were turning blue and she wanted him to take off his jacket? With one minute to air, the producer removed her parka, revealing shorts and a halter top.

"Lights!" she called.

Suddenly, Downs felt as if he was looking directly into the sun. It was hard to read the news copy, much less look into the camera. Within five minutes his clothing was soaked with sweat. As soon as the broadcast was over, the lights were switched off, the air conditioning went on and the producer again donned her parka. "You'll get used to it," she told Downs. He decided to stick with radio for a while but eventually filled a national anchor chair on NBC's Today Show and ABC's 20/20.[168]

TV newsmen struggled to make their presentations more visual. Jack Latham on KTLA in Los Angeles and Richard Hubbell on CBS-owned WCBW in New York were typical of 1940s news announcers who sat next to large maps. But with no studio monitors for reference, these well-intentioned newsmen sometimes pointed to the wrong location as they tried to maintain eye contact with the camera. A newsman at General Electric's WRGB in Schenectady, N.Y. tried to analyze the Pearl Harbor attack using toy boats and planes, only to have the strings become tangled. Announcers at Chicago's WGN in the early 1950s would hold up aircraft models to illustrate that "... this (Constellation) is the type of plane that just crashed. Except that the guy would pick up something else—a DC-3—and didn't know which end was which."[169]

Moviegoers of the 1930s and 40s were familiar with newsroom scenes of working print reporters and editors as characters in popular films.[170] To add the drama of news-in-the-making and give their television announcers more credibility, Chicago's WBKB put the first two-man news team in a newsroom setting. "Cubberly and Campbell" appeared in shirtsleeves with the sound of typewriters and ringing telephones in the background, but not because they were actually gathering the news themselves. General Manager William Eddy explained the "News Desk" format this way: "This double feature allows one actor to consider his lines as well as the picture. While his colleague is talking, there is an opportunity for him to develop well thought out questions and answers to sustain the interest of the audience." In addition to the two male "actors," WBKB also had a woman war correspondent on the staff. Ann Hunter thus became "not only an interesting commentator, but a particularly charming and acceptable picture as well." [171] However, this type of experimentation was rare. Typical local news broadcasts focused on a single male announcer at a desk.

***

Early audiences didn't mind if the announcer plugged a product in the middle of a news program. Advertising loomed large from the very first commercial TV news broadcast in 1941 on New York's WNBT. By contract, the distinguished radio newsman Lowell Thomas was upstaged by a stack of oil cans on the news desk. The sponsor's product was the only visual on the "Sunoco News."[172]

Gil Martyn, a former NBC radio newsman and voice of the Paramount Newsreel, wasn't above plugging Rancho Soups, the company that sponsored his newscast on KTLA. Reading the news with dramatic flair, the tall, dignified newsman would pause to sip soup and even comment on how good it tasted. Far from appearing tacky in 1948, the station considered it a prestigious accomplishment to have a sponsored news program. The soup-tasting was a nightly ritual on the most-watched evening news program in Los Angeles.[173] In the 1950s, Chicago's Fahey Flynn delivered the Windy City's top-rated news on WBBM while sharing the screen with a Standard Oil logo.

Some advertisements were delivered like news stories. Stan Chambers appeared on KTLA in a trench coat as a reporter "covering" comic attempts to promote Butternut Coffee with skywriting. Even the distinguished foreign correspondent Clete Roberts could not avoid plugging "the most dramatic success story in Buick's history" during a local TV news program. Beside him on the news desk was a pile of papers said to be press releases, glowing articles and telegrams from dealers, tall enough to hide an unidentified assistant standing behind it.[174] Stations who had weathermen used them as billboards for sponsoring products, typified by Philadelphia's Herb Clarke delivering his forecast on WCAU in the uniform of a Texaco gas station attendant.

TV stations hired news announcers with booming radio voices, but programmers began to notice that personality was just as important. NBC officials in New York noted a positive audience response when Lowell Thomas closed his broadcast with a wink. It made him "a member of the audience group rather than merely a picture traced in electrons."[175]

The importance of personality and good looks was even more evident on the West Coast. Actor Dick Lane, who appeared in over 250 films and frequently portrayed a journalist in the movies, was one of the first TV newsmen on KTLA. He also introduced televised wrestling to the nation as the announcer who made wrestling star "Gorgeous George" a household word in the 1940s. But because of his outstanding ability to ad-lib, the popular Lane continued to draw top news assignments.

Keith Hetherington, who had been a news announcer and disc jockey for KMPC radio, joined KTLA in 1945. Good looks were part of his appeal, according to TELE-views Magazine. "Tall, well-built and with a pleasant personality," the article said, "he is 41—and doesn't look it." Hired as a newsman, Hetherington also gained popularity as the original host of a live interview program, "Meet Me in Hollywood," and a chatty home shopping program, "Handy Hints." Owned by Paramount, a movie studio accustomed to public curiosity about its film stars, KTLA provided fan magazines with personal details about its TV personalities. Both Lane and Hetherington were promoted in the press as married men with children, exemplifying the family values of the post-World War II period.

But neither man was selected in a 1950 contest to name "the most attractive man on television today." Martha D. Jones of El Monte, California, won $5 for her prize-winning letter: "He is handy handsome and hospitable. His name is suggested at the mere sight of a television antenna. Children cry for him, men envy him, women love him! Who is he? Why he is the A-1 ad-libber, the charming gentleman M.C., the Robert Taylor of video—KTLA's own Personality Kid, Stan Chambers."[176]

Chambers achieved instant fame in 1949 when he was on the air for more than 27 continuous hours covering the attempted rescue of three-year-old Kathy Fiscus from an abandoned well. Ironically, his looks had nearly ended his TV career two years earlier. He had joined KTLA straight out of college in 1947. He was hired as a production assistant and helped build sets as he waited for his big break on the air. He got his chance during an interview with the Australian polo team. Snickering engineers couldn't help but comment on Stan's "kissable" lips. They looked fine in person, but through the lens of an old iconoscope camera, Stan looked like a male version of cartoon character Betty Boop. Better cameras fixed the problem with Chambers' appearance, but KTLA boss Klaus Landsberg still felt that the young broadcaster's voice was too high-pitched for TV.[177]

Chambers also struggled with an early version of the teleprompter in 1948. The awkward wooden podium was placed in front of the camera with rollers for a 50-foot scroll, like a giant roll of paper towels, containing the script in large type. "The trick was to place the top line as close to the lens as possible so that it looked as if the broadcaster were looking right into the camera lens," Chambers recalled. "I always knew there was the danger of the paper jamming, tearing or wrinkling at a bad angle for reading. Or the chance that the operator would get lost and not be able to find the place in the script that I was trying to read. I was terrified."

Chambers was one of the first newsmen to gain a foothold in TV without an extensive background in radio or newspaper journalism. Despite his relative inexperience and on-camera difficulties, he succeeded because of other qualities that would put the audience at ease. Landsberg believed viewers would shun pretentious on-air talent and would "much rather find a warm, friendly personality on the air that's considered one of them—one they welcome in their homes..." [178] Chambers' image as a regular guy was further enhanced by his hosting duties on an ice-skating variety show called "Frosty Frolics." Its finale usually called for the newsman-emcee to take a tumble onto the ice, surrounded by dancing skaters.

Viewers developed a personal relationship with Stan. If he read the news with a sniffle, they stopped by the studio with cold remedies. He was reporting on a beached whale in Santa Monica when a fly flew into his mouth. He continued his report while coughing and choking. "Mr. Chambers was so moved by the plight of the whale that he was in tears," an admiring viewer wrote to the station. "God bless him."[179]

What Barbara Matusow observed about network anchors in "The Evening Stars," also applied to local news announcers: "In the early days of television, the evening newscast was so primitive technically, so lacking in journalistic credibility, that the men who read the news were relatively humble figures. They were popular with the public, somewhat in the manner that game show hosts develop a following." [180] Indeed, Chambers was typical of many early newsmen who actually hosted game shows on the side, including Mike Wallace in New York and Bill Burns in Pittsburgh.

Inventing TV News

***

Despite the importance of looks and personality in a visual medium, early TV news announcers were also expected to have some journalistic ability. William Ray, the news director of the NBC station in Chicago, believed, "Television news requires the services of a newsman who can ad-lib his own show in front of the camera after selecting the news and writing his script in advance."[181] In the early 1950s, WNBQ's Clifton Utley in Chicago was typical of TV newsmen who were more than just copy readers. They wrote their own scripts and controlled the content of their programs. That didn't stop one critic from complaining that "you can't recall much about the news of the day, although you may remember all about Clifton Utley's haircut..."[182]

The lifting of the FCC's freeze on new licenses in 1953 meant that most sizable cities would have more than one TV station. A broadcast license to serve the public interest required all stations to make some effort in news. It kept the budget down if the anchorman could write his own copy and develop news sources to beat the competition. In Pittsburgh, KDKA's Burns gained a reputation for aggressive newsgathering. "He knew everybody in the city of Pittsburgh," said colleague George Eisenhauer, recalling how newsmakers would call Burns with story tips. "So he had a natural edge on everybody else in the news business." Burns also ventured out into the field with a film crew to shoot interviews, a rare initiative in the early days of bulky sound equipment. He would be the dominant local news figure in Pittsburgh for 40 years.[183]

However, the man who ruled Chicago's local news on WBBM from 1953 to 1968 didn't cover stories or write his own copy. Former radio announcer Fahey Flynn relied on a staff of writers, rehearsed with three cameras to perfect his delivery and mastered the use of the teleprompter. The genial Irishman greeted viewers with his trademark bow tie and a cheery "How do you do, ladies and gentlemen." WBBM's news director boasted that the staff could "write so well that you'd swear Flynn was ad-libbing."[184]

As late as 1960, a survey revealed that nearly all TV news personalities in Chicago had some type of background in radio or print journalism. "The trend that emerged was that of a combination of sound newsman and effective broadcast communicator," observed one Chicago broadcast historian. "'He sounds as if he knows what he's talking about' became evidence of success."

Whether he was producing his own material or relying on skilled copywriters, the image of the local news anchor had quickly fulfilled an early prediction that TV audiences would demand news presenters who at least appeared to possess superhuman qualities: "The television announcer must be well informed; he must be a quick thinker, nimble witted, and must choose words that fit the picture... he should be infinite in faculty, and in apprehension like a god."[185]

If this had been a help-wanted ad, George Putnam would have answered it.

Inventing TV News

***

Born in 1914, Putnam grew up in Minnesota. During the Depression of the 1930s, his father lost his job as a restaurant equipment sales manager and was reduced to selling peanuts door to door. His mother died of cancer. The family's financial difficulties forced young George to postpone his dreams of law school and go to work in 1934. On his 20th birthday, he got a job at 1,000-watt Minneapolis radio station WDGY. He earned $22.50 a week for duties that included washing the floors, answering the phones and spinning records. "I figured, I have a big mouth, I can make it in broadcasting," he recalled.[186] He was right. Putnam quickly moved on to a bigger station, WSTP in St. Paul. He decided to try his luck at the networks. In 1939 he competed with 20 other finalists and won a staff announcer position with NBC in New York. He did 14 programs a week, concentrating on news and special events. His 15-minute newscasts drew high ratings. He became one of the voices of the Hearst Movietone Newsreel and counted William Randolph Hearst, Lowell Thomas and Walter Winchell among his mentors.

"Walter Winchell made my career," he recalled years later. "The thing that gave me attention was he called me 'the greatest voice in American radio.' I went from $190 a month at NBC to better than $200,000 a year, which was too much for a 24-year-old kid out of Minnesota."[187]

During World War II Putnam served as a first lieutenant in the Marine Corps. He logged 20,000 miles on behalf of Armed Forces Radio. His post-war broadcasting career included appearances on the Mutual Broadcasting System, the BBC and the DuMont Television Network. He had married a female war correspondent, but the marriage dissolved into a nasty custody battle over the couple's four-year-old daughter. Tragically, the child died in the hospital after a tonsillectomy. "She was left alone and bled to death," Putnam remembered. "I fell apart. I just absolutely fell apart."[188]

The grieving father wanted a fresh start. He packed up his convertible and headed for Los Angeles, the back seat filled with his smartly tailored suits. A friend had promised him a job in TV, but by the time Putnam got to the West Coast the friend had been fired. He auditioned at KTTV and was hired as the station's newscaster in 1951.

Intense competition with KTLA led to the occasional embarrassing moment at the anchor desk. KTTV's response to Landsberg's fleet of mobile video units was to race around town shooting film, run it through the processor and rush to get it on the air. A newspaper article recalled the time that "Putnam breathlessly introduced 'unedited' footage of a brush fire, which proved a mistake, since the footage included a firefighter relieving himself at the edge of the blaze."[189]

Paying tribute to Putnam in 1984, newsman Bill Stout recalled a heart fund telethon in the early 1950s with close-up footage of surgery being performed on a young boy. The program cut to the boy's mother in the studio with Putnam, who said, "Mother, one thing we all can tell you, your boy has a lot of guts." [190] Such unintentional double-entendres only endeared Putnam to his audience.

More often, Putnam snared interviews with the famous and powerful, from Eleanor Roosevelt to Robert F. Kennedy and every U.S. President from Herbert Hoover to George Bush. When President Richard Nixon called Putnam for an update on the 1971 Sylmar earthquake, the anchor took the opportunity to question Nixon about U.S. policy in Cambodia. Israeli Prime Minister Golda Meir gave Putnam her recipe for chicken soup.

Putnam achieved the star status of having his name appear in the title of his broadcast, "George Putnam and the News." He quickly developed a following in Hollywood. He appeared as himself in movies, including "Fourteen Hours," which introduced Grace Kelly in 1951 and "I Want to Live," which won an Academy Award for Susan Hayward in 1958. His star on the Hollywood Walk of Fame is located between the stars for actor Mickey Rooney and actor-turned-politician Ronald Reagan.

Conditions on the West Coast were ideal for the local anchorman to adopt a celebrity lifestyle. Putnam drove fancy sports cars and dated movie starlets. Reminiscing on his 86th birthday with Hollywood columnist James Bacon, Putnam recalled their double date with Lucille Ball and Marie "the Body" McDonald. He felt secure enough in his position of authority to appear with Jack Benny in a 1959 comedy sketch as the host of a fictional talk show. He also developed a passion for horseracing, naming his ponies after his colleagues in the news business. Fans cheered when Putnam made his annual appearance in Pasadena's Tournament of Roses Parade, riding a magnificent Palomino with an ornate silver saddle.

Sam Zelman, a former newspaperman who was working for ABC-owned KECA in the 1950s, observed Putnam's rise in Los Angeles. Zelman said Putnam "introduced the flair and the personality cult into television in that he was very dramatic in his presentation." Zelman suggested that appearing on independent KTTV gave Putnam more freedom to experiment than his counterparts had at network-owned stations. "He developed a more conversational style than the style that was common at the time, which was newspaper style... cold-blooded sentences." Zelman, who wrote copy for future NBC star Chet Huntley, credited Putnam with writing "passionate sentences."[191]

Perhaps because of his celebrity status, Putnam was able to transcend the standard of objective reporting and publicly take sides, concluding his broadcast with a commentary called "One Reporter's Opinion." Putnam might sound off on the importance of protecting the elderly from abuse or rant against a Soviet plan to dump waste in space. He crusaded on behalf of Hungarian freedom fighters and Soviet Jews. "He always felt the responsibility of a television news reporter extends beyond the mere parrot-like recitation of the day's events," said Gary Owens, the Los Angeles radio announcer who would become the voice of the hit 1960's comedy show "Laugh-In." He said Putnam believed it was "his privilege and responsibility to right wrongs and to nudge the public conscience."

Putnam's advocacy was most notable in the 1961 Los Angeles mayor's race, which pitted incumbent Norris Poulson against City Councilman Pat McGee and businessman Sam

Yorty. Putnam defied the endorsement of the Los Angeles Times, which owned KTTV and backed Poulson. The anchorman disliked the incumbent and offered to put the two challengers on the news every night until the primary. McGee declined to appear after the first two nights, so Putnam continued the segment with just Yorty, giving the candidate one minute of free airtime each night for two weeks. Yorty advanced to a runoff and defeated Poulson.[192] The newspapers ran a victory photograph showing "Sam and George together in triumph, despite the opinions of his (Putnam's) employers."[193]

Management put up with Putnam's opinions because he drew a big enough audience to make his newscast profitable. At one point in the 1960s, his salary reached $350,000, making him the country's highest paid newscaster. "Walter Cronkite asked me what I made," Putnam recalled. "He told me he was making only $125,000."[194] After 14 years at KTTV, Putnam defected to rival KTLA in 1965. He returned to KTTV for more money three years later, finally landing at KTLA again in 1971. A 1984 Putnam tribute included humorous footage of the newsman repeatedly "crossing the street" between the neighboring rival stations, making more money each time. "George, you taught us all how to do that!" exclaimed KNBC's Kelly Lange, one of the first women to anchor a newscast in Los Angeles. "Follow the money," Putnam later joked.

"The guy was worth it because he had the ratings," said Hal Fishman, who teamed with Putnam at KTLA. "He brought in an enormous amount of money."

Terry Anzur

As a child of the Depression, Putnam took pride in his political heritage as a Roosevelt Democrat. He also set up a scholarship program for inner-city youth in the aftermath of the Watts riots. However, his pro-war stance in the 1960s riled audiences in the Vietnam era and caused Putnam to be labeled a conservative. "I've been called many things in my career," he told the Los Angeles Times. "... right-wing extremist, super-patriot, goose-stepping nationalist, jingoistic SOB. And those are some of the nice things!"[195]

But even Putnam was surprised when his larger-than-life persona provided the initial inspiration for an icon of popular culture.

***

America's best-known local television newsroom never really existed, except in the situation comedy world of the Mary Tyler Moore Show. Co-creators James L. Brooks and Allan Burns were on a tight deadline to come up with a weekly comedy series for Mary Tyler Moore, best known to audiences as the perky young wife on the Dick Van Dyke Show. The creative duo struggled with CBS executives who opposed casting Moore as a divorced woman on the new show, fearing that audiences would think the beloved Laura and Rob Petrie had split up. Divorce still carried a stigma in 1970, so the producers settled on the character of Mary Richards, a 30-year-old single, working woman. But what kind of job would she have? After discarding the idea of making her a researcher for a newspaper gossip columnist, they settled on a TV newsroom. This allowed the series to take viewers behind the scenes of a local newscast,

Inventing TV News

but it also required the writers to account for the lowly status of most women in television news at the time. In the first episode, Mary applies for a secretarial position that has been filled. She is hired as an associate producer, a job that pays $10 per week less than the clerical work. The news director says he intended to hire a man for the job and treats Mary in a condescending way as she struggles to discuss her qualifications.

There was also the question of geographic setting. According to Allan Burns, WJM-TV was located in Minneapolis because "the major industry is snow removal" and the creative team "wanted to trap the characters indoors. We wanted the audience to feel comfortable about a series that seldom ventured outside."[196]

Burns and Brooks only had to turn on their TV sets in Los Angeles to find a model for the character of the WJM anchorman, Ted Baxter. "It was going to be Putnam," Burns recalled. "He seemed to be such a pompous ass."[197]

Like Putnam, the character of Ted Baxter lacked a college education and got his start in radio. He would relish any opportunity to tell the story of his career—"It all started in a small 5,000-watt radio station in Fresno, California..."—only to be interrupted by the other characters in the newsroom who weren't interested in hearing about it. In one episode Baxter explained to a reporter that he got into television because God told him he was too handsome for radio. He covered the walls of his office with testimonials from "fans," including Pope John XXIII. "Popes watch the news," Ted explained to a disbelieving Mary and her friend, Rhoda.

Except for Brooks, who had worked in a TV newsroom in New York, the show's creative team needed more information to make the characters authentic. By lucky coincidence, Mary Tyler Moore's Aunt Bertie worked as the secretary to the general manager of KNXT, the CBS-owned station in Los Angeles. "Bertie gave us carte blanche to come over there and hang out," Burns said. "We watched all of the people in the newsroom and one of them was Jerry Dunphy. He struck us as being incredibly stupid, all the things that Putnam was except that he added the dividend of showing off all the time."

To the public, Jerry Dunphy was the formidable leading anchor on The Big News, the nation's first one-hour, local newscast launched by KNXT in 1960. The Big News was a team effort, including the pun-filled weather forecast by Bill Keene and sports with Gil Stratton. Dunphy was surrounded over the years with a strong cast of correspondents such as Joseph Benti, Bill Stout, Maury Green, Ruth Ashton-Taylor and Ralph Story. As Big News veteran Howard Gingold pointed out, "It was hard not to do a news broadcast that set standards of excellence" with "a peerless collection of producers and writers" to prepare Dunphy's scripts.[198]

By the time the writers for the Mary Tyler Moore Show arrived in 1970 to observe the news operation at KNXT, former CBS reporter Grant Holcomb had become the news director. According to Burns, Holcomb was the inspiration for the character played by Ed Asner, WJM news director Lou Grant. "He had this kind of beagle face," Burns explained, "and Jerry, knowing we were there, would show off by making Grant—that's where we got Mr. Grant—feel small."

Burns recalled one day when Dunphy complained about the thickness of the paper in the teleprompter. "He would say, 'Grant, this paper is too thin, get heavier paper.' He was always preening like a rooster, and he just gave us a lot of material. When Ted Knight read for us, he had that slightly vague, dumb quality."

Born Tadeusz Wladyslaw Konopka in 1923, Knight had done a little news announcing himself, on a small station back home in Connecticut. The distinguished-looking actor, who had played bit parts for years, was having trouble finding work in films when he read for the TV role. The producers had considered casting a younger actor as Ted Baxter, envisioning the handsome-but-vacuous anchor Brooks would later develop as the character played by William Hurt in the movie "Broadcast News." But the white-haired Knight, who physically resembled both Dunphy and Putnam, won the part.

"We still kept some elements of Putnam," Burns explained. "The two are not dissimilar, the physical thing with the white hair and *that voice*, but we never adopted his (Putnam's) right-wing views." As played by Knight, Ted Baxter was "a mass of insecurities. There was an episode where Mary discovered why Ted never took a vacation. They made him take a vacation and he said he was going to Mexico, but they got postcards that were postmarked in Minneapolis and he was watching the guy who subbed for him while he was gone."

***

Like Putnam and the fictional Baxter, Dunphy came from humble beginnings. Born in Milwaukee in 1921, he enlisted in the military prior to the U.S. entry into World War II because he needed the money. He served as an Air Force bombardier and studied journalism at the University of Wisconsin on the GI Bill. He began broadcasting at small radio stations, reading the farm news on WHA and WIBU in Madison. He moved up to a larger station in Davenport, Iowa, and became known as "Five-W Dunphy," promoting the journalism mantra of "who, what, when where and why" in his newscasts. But when he traveled to Los Angeles in the early 1950s and tried to make it as a radio announcer, the only work he could find was selling wholesale appliances as a manufacturer's representative. He returned to the Midwest, working in the sales department at his old station in Iowa until he was hired to organize a news department for a TV station in Peoria, Illinois. His next stop was anchoring at a financially troubled station in Wichita, Kansas. "It went into the toilet," he recalled, but Dunphy later in his career would point to this early hands-on experience as evidence that he was capable of doing more than just reading the teleprompter.[199]

Responding to an ad in a trade magazine, he auditioned for the CBS station in Milwaukee and won a job as a news announcer on his second try. But the big money was in Chicago. Dunphy couldn't compete with the established news stars in the Windy City, so he signed on as a CBS staff announcer. With a wife and three children to support, he did voice-overs for industrial films to supplement his broadcasting pay. "Those were very intense years," he would recall in an interview. "I was afraid of the embarrassment of failing. I didn't want to flop in front of my family."

Inventing TV News

Although he took pride in being able to write his own copy, Dunphy received less than glowing reviews. "There was a time when I was getting into broadcasting, particularly in radio, that they were so voice-conscious, that... if you didn't have stentorian tones, you didn't get the job," he said. "Chicago was the hub of the great voices in radio and... voice was the big thing. When I was getting started, I wasn't considered to have the kind of voice that would probably make it on a network. That has changed. My delivery has gotten better, practicing at public expense. I used to deliver out of nervousness, a lot more rapidly. When you do that, your voice is a little higher and good for sports... I never had the big voice for radio way back then and had I stayed in radio I would probably have failed." Fortunately, Dunphy found work in television as the sportscaster on Chicago's top-rated Standard Oil News alongside Fahey Flynn, the WBBM anchor who was legendary for not writing a single word of his program. Dunphy would later be accused of the same journalistic shortcoming.

By 1960, Dunphy's former boss from Milwaukee was in Los Angeles auditioning anchors for the program that would become The Big News. "They were even looking at actors and people who couldn't write. They knew I had run a news department and they knew how *this* went," Dunphy remembered, illustrating his point in Ted Baxter-ish fashion with his fingers pounding an imaginary typewriter. He signed on in Los Angeles to anchor a CBS regional 5 p.m. newscast and the 6 p.m. news on KNXT for a guarantee of $25,000 a year.

Like Putnam, Dunphy was able to fatten his paycheck by changing stations, moving from KNXT to KABC's Eyewitness News in 1975. His name never appeared in the title of his broadcast, but he was known for his signature greeting. According to Dunphy, it grew out of a conversation with KNXT news director Sam Zelman about whether to have a standard opening for The Big News. "We came up with something I could be comfortable with and that he didn't think was too pretentious: 'From the desert to the sea to all of Southern California, good evening.'" Dunphy was surprised when his new employers at KABC wanted him to keep using the same old line: "Everybody must be tired of hearing that. It gets tiresome for me. I wouldn't mind not doing it, wouldn't miss it."

Of course, Dunphy wasn't the only newscaster with his own personal catchphrase. Like Cronkite intoning "...and that's the way it is," many local anchors saved their signature lines for the end of the program. For George Putnam it was, "And that's the up-to-the-minute news. Up to the minute, that's all the news." For WTVJ's Ralph Renick in Miami, it was, "Good night and may the good news be yours." When Ted Baxter signed off with "Good night and good news," it was a parody of what viewers were hearing on their local channels.

Dunphy took pride in long workdays that took a toll on his family life. He claimed that he covered city hall with a film crew in the morning and then returned to KNXT to anchor the 6 p.m. and 11 p.m. newscasts. He boasted of reporting assignments that took him to the war-torn Middle East and the battlefields of Vietnam. He toured celebrity homes in a series called "Jerry Visits," and took a nostalgic trip to his ancestral home in Ireland. However, others behind the scenes of The Big News described some of Dunphy's adventures in the field as what might have happened if Ted Baxter had ventured outside the newsroom with a camera crew.

"No matter how many times (Dunphy) was told he was the best reader in the business, he wanted to prove to everyone that he was also one of the best journalists," said Joe Saltzman, a writer/producer at KNXT. He recalled the time Dunphy was sent to interview Jayne Mansfield, a shapely, blonde movie star. The film crew resented being bossed around by the anchor, but they dutifully executed his orders to set up the equipment next to Mansfield's swimming pool. Dunphy rehearsed a scenario that would have the actress jump into the pool and swim across to the microphone to be interviewed. The camera rolled. Everything was fine, until Mansfield paddled across the pool and mistakenly grabbed the microphone instead of the pool railing. "Something went wrong, some kind of electrical shock," Saltzman said. Dunphy "shot up into the air and flopped into the water... Every agonizing second of the Talent and the Star's fight to get out of the pool was recorded for posterity. There was no story on the Six O'Clock News that night. And the Great Reader didn't want to go out into the field again."[200]

Dunphy later maintained that his workload was reduced because of concerns about his health. He also blamed union regulations for discouraging his efforts to write more of the broadcast:

> "The union system took away from the anchors the right to do what I always did: walk in, pick up the copy and say, 'I'll write this.' But no, somebody is covering this and that goes to this writer because (the story) was covered by this reporter and the writer, under the union rules, would have to edit the piece for this reporter. It finally came down to where anchors couldn't touch the copy... If you are discouraged over a period of years, you would do what I did and say the hell with it... Did they discourage us or have rules and regulations that isolated the anchors? They sure as hell did. And they got what they asked for, not as much production out of the anchors."[201]

In fairness to Dunphy, the increasing complexity of television newscasts in the 1960s made it virtually impossible for anchors to control all of the content. "The expanded format meant the end of the newscaster writing his entire broadcast," Nielsen noted of the period between 1965 and 1968 in Chicago.[202] On the network level, even the esteemed Walter Cronkite rarely wrote his own scripts and functioned as more of a managerial copy editor.[203] However, Dunphy did not help his own reputation with his support staff by eagerly pounding out copy during a writers' strike at KABC and generally downplaying the contributions of those who toiled behind the scenes.

Saltzman recalled Dunphy telling a news writer, "You could be on the air if you got a nose job, did something about that complexion, got a new hairstyle, learned how to dress and took voice lessons." The anchorman reportedly went on to criticize a zoning story, telling the writer it didn't make sense. "Either shape up or we'll get someone else."

"We're trying to get out a newscast here," the writer replied. "Don't you have something to do?"

"Yeah, I have plenty to do," Dunphy retorted. "Without me, you're nothing. Without me you might as well go home. I'm the reason everybody tunes into the news. So who needs you? Without me you're just a piece of blank paper."

The writer, who produced Dunphy's 11 p.m. newscast on KNXT, plotted revenge.

The next night, the entire writing staff was told to sit at their desks and do nothing. The director of the broadcast was told to go to the control room—without a script—and keep his mouth shut. At 10:15, Dunphy called the writer to ask for a script and was told to go f--- himself. According to Saltzman, the anchor became more and more agitated as the minutes to airtime ticked by. He called the news director, who wasn't home. He called the general manager, who called the writer and demanded to know the whereabouts of Dunphy's 11 p.m. script. He got the same profane answer that Dunphy had received. The anchor stalked into the studio and threw the makeup man's powder puff across the room. At 10:59, Dunphy was on the set, pale and shaken. He had no script and was told the teleprompter was not working. He was about to go live in front of an audience of three million people. In Dunphy's words, he was "out there, naked."

Meanwhile, the writer had secretly produced the entire program ahead of time. The script had been hidden in his desk drawer. At the last minute he distributed copies to the technical crew but sent only one page out to the set for Dunphy. Saltzman recalled that the script was delivered to the anchor desk page by page "to the frantic talent who grabs the new page seconds before he finishes the page he is reading." Somehow, Dunphy got through the show and returned to the newsroom to find the offending producer with his feet propped up on his desk, enjoying a beer. "Nice job," he yelled to Dunphy. The next day, the anchorman called in sick.[204]

Three months later, the news director remarked on how well Dunphy and the producer were getting along. "I've never seen such mutual respect in this business," the news director was heard to say. It could have been an episode of the Mary Tyler Moore Show, with its humorous tension between Ted Baxter and news writer Murray Slaughter.

"The night Murray found out just how much money Ted Baxter was making was an episode that hit home," Saltzman wrote. "Ted Baxter epitomizes the worst of every anchorman and sums up verities about the business from the writer-producer's standpoint: anchormen can be irrational, stupid, silly, obstinate, ridiculous and get away with it. If a writer makes a mistake, no matter how minor, he is chewed out by everyone from the news director on down. He's only as good as his last paragraph."

Like Putnam, Dunphy played himself as a newscaster in the movies and made headlines with his personal life. While driving in his Rolls Royce, Dunphy and his then-fiancée were shot and wounded in what police called a botched robbery attempt on October 24, 1983. "If I hadn't been able to afford a Rolls Royce then I would have been driving a less conspicuous car and the crumb-bums that did this maybe wouldn't have turned around and attacked us," Dunphy said later. "Did my fame almost contribute to my demise? I was not shot because I was Jerry Dunphy. I looked like a fat cat in a fancy car that they anticipated robbing."

Dunphy insisted that news anchors should not seek celebrity status. "The business seems to have made us that. I think anchors wanting to start out to be stars because they are anchors wouldn't work that way," he declared. "Cronkite became a star news guy because he was star quality, a super excellent man who knew the news business and how to do it as well as anybody. If that made him a star, so be it. I thought he was just a damn good newsbody myself and can't help it if you're perceived that way." Dunphy perceived anchor stardom to be a function of corporate efforts to promote local and national news as a product. "I think the hype for Dan Rather when he was taking over for Cronkite was to make him a star. And he was just a damn good reporter who deserved the job. But did CBS not hype Dan Rather? You bet your butt they did."

Dunphy volunteered a comparison between anchors like himself and one of the most beloved movie actors of all time. "You can't make yourself a star, no more than Jimmy Stewart would have been a star if somebody didn't promote him. I knew him when he was a fair to middling pilot on training missions in Albuquerque, New Mexico, and he used to get lost over the desert."

Management put up with Dunphy's ego because he was popular with viewers. Even his harshest critics could not deny his ability to communicate with the audience. "Working from a prepared script or from a hurriedly scribbled bulletin, there are few to equal him," Saltzman acknowledged. "As long as it was an affair between Dunphy and the written word, he was without peer."

An unidentified KNXT producer said of Dunphy, "A lot of people condemned him because he simply read other people's copy, but goddamnit, he was the best reader this city ever had. He was what every anonymous writer needs—the best possible presentation your copy could get, life and believability breathed into your words."

Nobody ever said that about Ted Baxter.

***

The Ted Baxter character created problems for the actor who became closely identified with the role. Knight complained to the producers, "I can't do this. I can't play this character, this stupid arrogant, ignorant man who is a laughingstock. It's just gotten into my soul... I'm identified with this person and I just don't want to do it anymore."

Realizing that "our biggest laughs come from him," Burns reassured the actor by comparing the role to the classic comedy of Jack Benny, who appeared to be cheap and conceited but was still popular with the audience. "Gradually, I think, (Knight) realized that people really loved the character."[205]

But the real-life models for the role of Ted Baxter didn't like it one bit, a reaction that surprised Burns. "I thought at the time George Putnam would not recognize himself in the character," he said.

"Early on, it offended both Jerry (Dunphy) and myself," Putnam told the Los Angeles Times. "Because when you're much younger you take yourself more seriously. Then you outgrow it and you don't give a damn. It was a hell of a cute character. And I've known a few (anchors) like that. Airheads."[206]

Local television news would ultimately embrace Dunphy, the objective news reader, and reject the opinionated commentator personified by Putnam, who fell from grace in the early 1970s. "I wore out my welcome," he acknowledged in an interview.[207]

In 1971, Putnam presided over a confrontational hybrid of news and opinion called "Talk Back" on KTLA, an early experiment with interactive news. The hour-long newscast began with 30 minutes of traditional coverage read by Putnam, Fishman and Larry McCormick, one of the first African Americans to anchor in Los Angeles. Then in the second half-hour, the cameras would pivot to focus on a studio audience, asking questions and debating issues with the anchors and invited guests. A typical face-off pitted the Jewish Defense League against members of the American Nazi party. The nightly drama was popular at first, but when the novelty faded, so did the ratings. Paramount had sold KTLA to singing cowboy Gene Autry, who was persuaded to drop Putnam in favor of the less controversial Fishman. At Christmas in 1973, Putnam's contract was not renewed.

His fate was typical of TV news icons around the country being edged out by younger, more energetic newsmen. But the entrenched anchors didn't give up without a fight. In

Philadelphia, John Facenda dominated the news on WCAU, the CBS-owned station. Most Americans would recognize Facenda's powerful bass delivery as the voice of highlight films for the National Football League. He rivaled Vince Leonard at KYW for ratings supremacy in the City of Brotherly Love. But in 1970, the ABC affiliate, WPVI, decided to compete by introducing a flashy new format called "Action News." Its temporary anchor was a young radio reporter from Miami named Larry Kane. In his memoirs, Kane recalled what happened when Facenda invited him to dinner: "Young man," said Facenda, "our CBS station in St. Louis, KMOX, is looking for a five o'clock anchor and they are very interested in you. It would be a wonderful opportunity and I think you should fly out and talk to them."

Kane was flattered but later realized "that the great John Facenda's mission was really to get me out of town." Kane became the permanent anchor at WPVI. Two years later, Facenda lost his anchor position as WCAU changed its format to compete with the faster-paced Action News. "I still believe that was a gross mistake on the part of WCAU management," Kane said. "Facenda was loved; removing him guaranteed more than a decade of rating and financial losses at the station John Facenda had built."[208]

Other stations paid the price when their elder statesmen crossed over to rival stations. Fahey Flynn was replaced at WBBM in Chicago by Bill Kurtis, a rising star from Kansas. The popular Irishman defected to WABC in 1968, becoming the cornerstone for the "Eyewitness News" team and taking the audience with him. Flynn's trademark bow tie "looks great on Channel 7," one admirer wrote to the station. "Long may it wave."[209] Flynn was a dominant figure in Chicago news ratings throughout the 1970s and died at age 67 in 1984.

In Los Angeles, Dunphy also defected to the "Eyewitness News" format in 1975. Former KNXT producer Gerald Ruben relished telling the tale of how KABC general manager John Severino stole the CBS station's star by trash-talking during a tennis match with the general manager from Channel 2. Dunphy's contract was about to expire and Severino voiced the opinion that The Big News veteran was too old for the marketplace. Dunphy was fired from KNXT and was drowning his sorrows at a local bar when the phone rang. It was Severino, calling with a lucrative contract offer to join Channel 7.[210]

Eyewitness News would battle KNBC for number-one ratings for the rest of the century, while Channel 2 would struggle in vain to recapture the big numbers associated with Dunphy and The Big News. In 1989, he was paid $1 million a year to anchor a three-hour prime time news block on Disney-owned independent KCAL. By the time he returned to KCBS in 1995, it was too late for even Dunphy to restore the glory days of The Big News. He returned to KCAL to read the 9 p.m. news.

"It's easy to make fun of Dunphy... the inspiration for Ted Baxter, the not-so-smart and all-too pompous anchor of the Mary Tyler Moore Show," observed Los Angeles Times columnist Bill Boyarsky. "He's had lots of career setbacks, but Dunphy's Irish face and thick white hair have been trademarks of L.A. TV news through the medium's many transformations and he's still an anchor. In short, Dunphy is a survivor..."[211]

Putnam also survived, but without compromising his opinionated style. He co-hosted an Emmy-winning show with satirist Mort Sahl and continued his "often imitated, never duplicated, original Talk Back" program on a 20-thousand watt AM radio station, drawing a small but faithful audience for more than 25 years. Resentful over losing his TV status, he mourned what local newscasters had become. "No news reader is worth $1 million on a local station," he said, complaining that successful anchors no longer were able to write. "I would like to ask the average broadcaster today after he or she is finished broadcasting to repeat for me what he or she just said. Now there would be a test! And I'll wager there isn't one in ten who has absorbed the material because they didn't write it, prepare it or report it."

In 1984, Knight hosted a KTTV tribute to George Putnam. In the format of a Hollywood celebrity roast, the newsman took a friendly ribbing from such entertainment legends as George Burns, Bob Hope and Sammy Davis Junior. Ed Asner told Putnam, "Some people used to say that the Ted Baxter character on the Mary Tyler Moore Show was based on you. That's ridiculous. Baxter was much better at reading the teleprompter." The audience howled as Knight rubbed his eyebrows, imitating Putnam, who responded with his own imitation of Ted Baxter. Later came a tape of Jerry Dunphy warning Putnam to "quit doing Ted Knight" because "he does our act."[212] All three men appeared to have come to terms with their shared public image. They were in on the joke.

Ted Baxter became "an indelible image in the public's mind of the pompous, perfectly groomed, know-nothing local news anchor," Saltzman observed.[213] The rest of the cast presented "some of the most positive images of journalists in the history of movies and television. These journalists are people we would like to spend a half hour with every week, people we would like to have over for dinner or a cup of coffee."

The show also represented a throwback to a simpler time in TV news when the producer's biggest problem was whether the anchorman would stumble over the copy. The stakes got much bigger when outside consultants and corporate bean-counters began meddling in local TV news content. The 1976 film "Network" was much more relevant to what was happening as real TV newsrooms adopted entertainment values. The ambitious programmer, played by Faye Dunaway, argues that an aging anchorman should become "the mad prophet of the airwaves" because "it looks like he may go over bigger than Mary Tyler Moore."

The situation comedy aired from 1970 to 1977, years when many real stations were promoting a warm and friendly image for their news teams. Viewers had no trouble making the connection. Burns cited an episode called "The Good Time News" as an example: "Everybody was going into happy talk and they (the people at WJM) realized they couldn't possibly do it with a lummox like Ted. They put him with a weatherman played by John Amos who had charisma and Ted got nervous... Of course, Ted was upstaged all the time. Mary had to come on to do an editorial because Lou didn't want to do it. And Ted tried to tell jokes. That's when Mary tells him to shut up. It was very funny." Historians of the show noted that this plot was "an idea whose time had seemed to come in the early seventies and there was much supercilious discussion in the print media about the idiocy of hundreds of grinning and joking anchor men and women all over America."[214]

The later years of the Mary Tyler Moore Show reflected the rise of women and minorities. Mary Richards and African-American weatherman Gordie Howard received job promotions. Lou Grant and Murray Slaughter represented a generation of TV pioneers with roots in newspapers or radio, replaced by young managers who grew up watching the news on television. The characters also felt the influence of hired consultants. New management fired everyone on the staff—except Ted Baxter. The anchorman was the only survivor at WJM-TV when Mary Richards turned out the lights in the final episode. It symbolized the end of an era for local television news as a money-losing public service and the beginning of its transformation into an entertainment-driven profit center.

Chapter Six:
The Rest of the Story

Of course, there's a lot more to the story.

I had hoped to go on from here. The next chapter would have examined local TV coverage of events leading up to the 1965 civil disturbance known as the Watts Riots and the live broadcasts on Los Angeles TV stations during the unrest itself. In the aftermath, TV newsrooms would have to confront questions of race and inclusion. The mostly white and male news departments would need to become more diverse. Not only because it was the right thing to do, but because of FCC mandates for more people of color and women in both staffing and news content. Over the next decade, progress was slow. A 1977 report by the U.S. Commission on Civil Rights concluded that affirmative action guidelines had produced mere "window dressing on the set."

That's where my own story becomes part of the narrative. Stations reached out to print outlets and radio stations to recruit more women and minorities with journalism experience. I was a 23-year-old Stanford University graduate covering the Rhode Island state capitol for United Press International. A local TV station asked if I'd give television reporting a try. I joined WPRI-TV in Providence in time to cover the Great Blizzard of 1978. It was only a few months before other stations came calling with offers for me to report and anchor in bigger markets. Twenty years and many moves later, I had achieved my dream job of co-anchoring KTLA News at Ten—the station I had watched as a kid growing up in suburban Altadena.

In 1997, I made a small contribution to KTLA's history by being the first woman to say her name at the top of the signature 10 p.m. newscast. Women had co-anchored the program for years, but were referred to by crew members as the "human cough buttons." Anchorman Hal Fishman believed that he was the main reason viewers tuned in. He would read the first three or four stories, with the female sidekick being assigned to speak only when Hal was at risk of running out of breath. He grudgingly went along with a management edict that the program would begin with both of us greeting the audience, something that had long been standard practice for news teams all over the country.

During my three years on the program, KTLA News at Ten won awards and regained the number-one position in the ratings. However, my contract was not renewed. Absent any other explanation, I concluded that Hal apparently didn't want to share the credit. Behind the scenes, it had long been said of Mr. Fishman that "he changed his toupee and his co-anchor every three years."

At the same time, I had been unable to find a publisher interested in local TV history. It wasn't "serious" enough for the academic press. Commercial publishers said I wasn't enough of a big-name author and asked if I'd be willing to ghost-write it for someone more well-known. Without a published best-seller, my chances of gaining tenure at USC

Inventing TV News

Annenberg in 2001 were slim to none. I moved on to another TV anchoring job and ended up starting my own media training company a few years later. Once again, the early history of local TV news in Los Angeles ended up in a file cabinet—mine.

I'd like to thank my son, Andrew Anzur Clement, PhD, for putting this book out there. Andrew was the school kid who loved to come to the station to join mom for dinner break and watch her do the news. He took my advice to not become a journalist, although he might have been a good one. Instead, he became an accomplished political scientist and author of more than two dozen works of fiction. TV reporters are often central characters in his stories. I'm grateful for his support and his e-publishing skills.

This book is also an acknowledgement of the many KTLA colleagues who reminisced with me about their early days in the business. In particular, I'd like to thank John Silva, Cleve Landsberg and Evelyn DeWolfe for consenting to be interviewed. Alice Fiscus graciously agreed to answer my questions in writing, and Clyde Harp provided the sandhogs' point of view on the rescue effort.

During my time at USC, professors Joe Saltzman, Ed Guthman, Ed Cray and Jack Langguth never wavered in their support for this research or my teaching skills. Bill Knowles of the University of Montana encouraged me to present this research for peer review and discussion. I will never forget these trusted mentors and friends.

I would never have survived as a working mom with two full time jobs without the support of my parents, Ed and Ellie, and my husband, Bill.

I hope this book will bring a smile to the faces of "news geezers" everywhere. Perhaps other historians or journalism students will find it useful. I still hope that the story will someday be told to a wider audience on a bigger screen. I even wrote a screenplay about the attempted rescue of Kathy Fiscus, the little girl whose tragic death made such a big impact on the evolution of the news we consume today.

Thank you for traveling back in time with me and taking the time to leave a review. As Ralph Renick at WTVJ in Miami used to say, may the good news be yours.

Citations and Endnotes

[1] Joe Saltzman, "Coast to Coast," *The Hollywood Reporter*, 4 June 1975. KTLA archives.

[2] Patt Morrison, "The Little Girl Who Changed Television Forever," *Los Angeles Times Magazine*, 31 January 1999, p. 9.

[3] Alice Fiscus, correspondence with author. All quotes from Alice Fiscus in this chapter are from this source unless noted otherwise. She disputes accounts that Kathy intentionally climbed into the hole while playing hide and seek.

[4] Bill Johnston, "The Kathy Fiscus Tragedy," *Los Angeles Daily Commerce*, 27 October 1987. KTLA archives. All Johnston quotes from this source unless noted otherwise.

[5] Elaine St. Johns, "Kathy Knew Not Horror, Nor Pain," *Los Angeles Mirror*, 11 April 1949, p. 3. Both the *Mirror* and *Los Angeles Times* had extensive coverage of the Fiscus story on this date, providing the factual basis for this narrative unless noted otherwise.

[6] Evelyn DeWolfe, "The Day Live TV News Coverage Was Born," *Los Angeles Times*, 17 October 1987. Calendar Section, p. 1.

[7] Dan Jenkins, "KTLA... the station... and the man," *TELE-views*, 20 January 1950.

[8] Evelyn DeWolfe Nadel interview.

[9] "Garage TV and How It Grew," Program notes from 52nd Annual Los Angeles Area Academy Awards, 17 June 2000, pp. 24-25.

[10] Cecil Smith, "The TV Scene: Born Yesterday? No, Just Retarded," *Los Angeles Times*, 22 January 1962. KTLA archives.

[11] Celia Rasmussen, "Deadly Blast a Proving Ground for Live TV," *Los Angeles Times*. Undated clipping from KTLA archives.

[12] "KTLA's Transmitter Set the Southland Pace," undated mid-1950s press release, KTLA archives.

[13] This was a widely published estimate; a Broadcasting survey estimated over 100,000.

[14] Descriptions of Welsh and Chambers from "KTLA… the team," TELE-Views, 20 January 1950, p. 11.

[15] John Silva interview with the author.

[16] Stan Chambers, *News at Ten: Fifty Years with Stan Chambers*, (Santa Barbara, California: Capra Press, 1994), 52-62. Chambers account of rescue coverage is from this source unless noted otherwise.

[17] Chambers and Welsh gave this account during remarks at the "Remembering Kathy" memorial attended by the author, San Marino, California, 11 April 1999.

[18] Stan Chambers, "The Kathy Fiscus Telecast Forty Years Ago This Weekend," undated manuscript in KTLA archives, p. 4. This passage apparently was cut from a published version of this article in the *Los Angeles Times,* 8 April 1989, Calendar Section, p. 1.

[19] Clyde Harp interview and correspondence.

[20] Smith, op. cit.

[21] George Putnam in *KTTV 35th Anniversary Show*, aired 1 January 1984, video recording in UCLA Film and TV archive.

[22] Hatch, Anthony, "Television's Baptism of Fire," *Los Angeles Times*, Last Page, 8 April 1979, KTLA archives.

[23] This quote and description of this phase of the rescue from "Five Brave Men Toil Deep in Shaft," *Los Angeles Times*, and supported by "Grimy Guys were Real Heroes," in the *Mirror*, 11 April 1949, KTLA archives.

[24] Harp correspondence.

[25] Bill Welsh, interview with Tom Henenkius for unpublished student project, University of Southern California, manuscript e-mailed to author 6 June 2000.

[26] Judy Farah, "Current events recall past tragedy," Associated Press, 16 October 1987. Clipping from *Pomona Progress Bulletin* in KTLA archives.

[27] Partial recording of Fiscus coverage, quoted in Mark J. Williams, *"Remote" Possibilities to Entertaining "Difference": A Regional Study of the Rise of the Television Industry in Los Angeles, 1930-1952*, (University of Southern California dissertation, 1992), 173-174.

[28] Welsh, "Television Grows Up," *Preview* Magazine, 15 August-September 1985, KTLA archives.

[29] Harp correspondence.

[30] Morrison, op. cit.

[31] DeWolfe, op. cit.

[32] Arthur Unger, "A Plaque for KTLA," *Daily Variety*, 12 April 1949, quoted in Chambers, *News at Ten*, p. 64 and Williams, p. 123.

[33] "Television has 27-hour Fire Trial," *Los Angeles Times*, 11 April 1949, p. 2.

[34] Dan Jenkins, "KTLA... the Station," *TELE-Views*, 20 January 1950, p. 6.

[35] "50 Years of Television Retailing," *Dealerscope*, Vol. 19 No. 4, April 1977, pp. 18-19.

[36] St. Johns, op. cit.

[37] "One Little Girl," *New York Times*, 11 April 1949, p. 24.

[38] Welsh, "Television Grows Up," op. cit.

[39] Maggie Chambers interview with the author.

[40] Chambers, *News at Ten*, op. cit., pp. 62-63.

[41] Elizabeth Lee, "Remembering Kathy," 10 April 1999, *Pasadena Star News*, Celebrations section, p. 1.

[42] Walter Ames, "Death Case Coverage Brings Drama to Viewers," *Los Angeles Times,* 26 May 1951.

[43] Eric Pooley, "Grins, Gore and Videotape," *New York Magazine,* Vol. 22, No. 40, 9 October 1989.

[44] Quoted in Hal Humphrey, "TV Covers First Crime Hunt," *Los Angeles Mirror*, 24 May 1951, KTLA archives.

[45] Susan Wilbur, *The History of Television in Los Angeles: 1931-52*, pp. 10-13. Master's Thesis, University of Southern California, January 1976.

[46] Harry R. Lubcke, "Television on the West Coast," John Porterfield and Kay Reynolds (eds.), *We Present Television,* (New York: W.W. Norton, 1940), p. 228.

[47] W6XAO was commercially licensed as KTSL in 1948, sold to CBS in 1951, renamed KNXT and finally KCBS.

[48] Michael Ritchie, *Please Stand By: A Prehistory of Television*, (Woodstock, NY: Overlook Press, 1995), p. 124. The first network gavel-to-gavel convention coverage of the political conventions was in 1948 and "inspired the networks to get serious about the possibility of news reporting on television." However, it could not be

seen nationwide. The first truly national convention coverage was made possible in 1952 because of the relay installed the previous year.

[49] Lynn Boyd Hinds, *Broadcasting the Local News, The Early Years of Pittsburg"s KDKA-TV*, (University Park: University of Pennsylvania Press, 1995), p. 21.

[50] "KTLA Station Operations," article reprinted from *Television Magazine* in KTLA Archives, undated but probably 1947 because it includes the station's "recently issued rate card" with fewer than 1,000 sets in use.

[51] Theodore Lynn Nielsen, *A History of Chicago Television News Presentation 1948-1968,* The University of Wisconsin, dissertation, p. 45.

[52] For details on "Pitt Parade" see Hinds, op. cit., pp. 35-60.

[53] Communications Act of 1934, Washington D.C. Wilbur, op. cit., pp. 28-30 also cites the FCC's 1946 "blue book" as an attempt to enforce the public service requirement.

[54] Raymond Fielding, *The American Newsreel 1911-1967*, (Norman: University of Oklahoma Press, 1972), p. 3. Cited in Hinds, p. 43.

[55] Chambers, *News at Ten*, p. 47.

[56] Gary Cummings, "The Watershed in Local TV News," *Gannett Center Journal* (1987).

[57] Chambers, op. cit., pp. 90-92.

[58] John Silva interview with the author.

[59] Williams, pp. 83-84, discusses the concerns of regulators in this area.

[60] Sig Mickelson, *The Decade that Shaped Television News: CBS in the 1950s*, (Westport, CT: Praeger, 1998), p. 23.

[61] Nielsen, op. cit., pp. 69-73.

[62] "KTLA… in the public interest," *TELE-Views*, 20 January 1950, p. 9.

[63] The account of the Patty Jean Hull kidnapping is drawn from newspaper reports in the *Los Angeles Examiner*, 22 May 1951, and the *Los Angeles Times*, 21-27 May 1951 unless noted otherwise.

[64] Chambers, op. cit., pp. 149-150.

[65] KTLA press release, 25 May 1951.

[66] Humphrey, op. cit.

[67] "Papers Squawk as TV Racks Up Beats," *Hollywood Reporter*, 24 May 1951. KTLA archives.

[68] KTLA press release, op. cit.

[69] Stan Chambers interview with the author.

[70] "TV Coverage of Tot Tragedy Grips LA," *Variety*, 25 May 1951, KTLA archives.

[71] Humphrey, op. cit.

[72] KTLA press release, op cit.

[73] Barney Glazer, 20th Century News syndicated column as printed in the *Beverly Hills Citizen*, 1 June 1951, clipping from KTLA archives.

[74] Wilbur, op. cit, pp. 148-149.

[75] Williams, op. cit. p. 102.

[76] "KTTV Newsreel to show Southland Men at War," *Los Angeles Times*, 26 May 1951.

[77] Chambers, op. cit., pp. 80-84.

[78] Los Angeles TV stations were spared the "Red Channels" hysteria in New York. Only CBS required employees to take a loyalty oath.

[79] Account of hearings based on reports in the *Los Angeles Times*, 18-23 September 1951. It should be noted the Times continued the

self-serving practice of reporting only on the activities of its own station, KTTV.

[80] "3 Claim Credit for Probe Okay on TV," *Variety*, 19 September 1951, KTLA archives.

[81] Letter from 10th District California Congress of Parents and Teachers, quoted in the *Los Angeles Times*, 20 September 1951. There is no known recording of these broadcasts and KTLA personnel claimed to remember little of the content.

[82] Frank Orme, "KTLA Sets Up Model Service Schedule," *TV Magazine*, undated 1951 clipping, KTLA archives.

[83] Jeff Greenfield, "Making TV News Pay," *Gannett Center Journal*, Spring 1987, p. 29.

[84] Klaus Landsberg, "Knowing the Pulse of Your TV Audience," in *Twenty Two Television Talks,* (New York: Broadcast Music, Inc., 1953), pp. 136-152. Cited in Williams, p. 182.

[85] "KTLA's 'City at Night' Can Sidetrack Santa Fe," Variety, 16 February 1951, p. 6.

[86] Williams, p. 180.

[87] Evelyn DeWolfe Nadel interview with the author.

[88] "Blast of Atom Bomb Thrills Television Audience," *Los Angeles Times*, 7 February 1951.

[89] Williams, pp. 216-218.

[90] Unless otherwise noted, the account of the atomic bomb broadcasts is drawn from the actual recording in the Museum of Radio and Television, and: Chambers, op. cit. p. 84-90, and "They Said It Couldn't Be Done: The Story of the First Televising of an Atomic Detonation," remarks of Charter Heslep, AEC Public

Information Service, and quoted in *KTLA West Coast Pioneer*, Museum of Radio and Television catalog, 1985.

[91] John Polich interview, *KTLA at 20*, videotape from KTLA archives. All Polich quotes from this source.

[92] "Washington Viewers Unimpressed by TV," Copley News service wire story, 23 April 1952, clipping in KTLA archives.

[93] Quoted in Heslep, "The A-bomb comes to America's living rooms, live," *The Quill*, February 1988, pp. 35-37.

[94] Sacramento Bee cartoon reprinted in "*KTLA West Coast Pioneer*," op. cit., pp. 38-39.

[95] Hal Boyle, "Atomic Blast Telecast Shows America Has Lost Early Fears," Associated Press, undated clipping in KTLA archives.

[96] John C. Mahoney, "Setting the TV Record Straight," Performing Arts, December 1972, p. 19.

[97] Barbara Matusow, *The Evening Stars*, (New York: Ballantine Books, 1983), p. 55.

[98] Williams, op. cit., p. 107.

[99] Nielsen, op. cit., p. 80-81.

[100] Humphrey, "A-Bomb: Courtesy of Klaus," *Los Angeles Mirror*, undated clipping in KTLA archives.

[101] Ann Terrill, "KTLA, Channel 5: Western Pioneer," *The Register* Leisuretime, TV Log, 28 May 1967, p. 11.

[102] *KTLA at 20*, op. cit.

[103] "Khrushchev Visits Can-Can," KTLA video recording, Museum of Radio and Television.

[104] Due to a rule that prohibited cameras above the level of Khrushchev's head, the Telecopter was not used.

[105] Barney Glazer, "Televents of the Week," *Highland Park News Herald*, 23 January 1953. KTLA archives.

[106] "KTLA Special Events Put Viewers at Scene." Undated 1950s KTLA publicity materials, KTLA archives, p. 22

[107] *Los Angeles Herald Examiner*, Friday, 16 January 1953. Page A-9. Picture shows KTLA camera just a few feet away from pounding waves.

[108] Glazer, op. cit.

[109] *Redondo Beach Daily Breeze*, 15 January 1953. Account of rescue and drenched officials.

[110] The Hollywood Reporter, 16 January 1953.

[111] Paramount Television Productions, Interoffice communication. 15 January 1953, in KTLA archives.

[112] Glazer, op. cit.

[113] "KNBH Scores a Beat On Quake Coverage; KTLA first on live," *Variety*, 22 July 1952. KTLA archives.

[114] "Cleve Landsberg's Recollection of KTLA and His Father," in *KTLA West Coast Pioneer*, Museum of Broadcasting catalog, 1984, p. 33.

[115] Chambers, *News at Ten*, pp. 88-90.

[116] Walter Ames, "Filmed Video Scores Quake Beat," *Los Angeles Times*, 23 July 1952, KTLA archives.

[117] All John Silva quotes are from a 2000 interview with the author unless otherwise noted.

[118] John Silva biography, "Telecopter News," undated KTLA Paramount Television Productions press release from KTLA archives.

[119] Ritchie, *Please Stand By*, p. 28.

[120] Chambers, op. cit. pp. 103-105.

[121] Thomas Hutchinson, *Here is Television,* (New York: Hastings House, 1946), p. 219. Cited in Nielsen, p. 29.

[122] Chambers, *News at Ten*, pp. 116-119.

[123] Silva memo to Jim Schulke, private collection of John Silva.

[124] "Now Comes the Telecopter," *Variety*, 30 July 1958. KTLA archives.

[125] "L.A. Scores Another First," *TV-Radio Life*, 13 September 1958. KTLA archives.

[126] "The TV Scene: Whirlybird New KTLA Reporter," Cecil Smith, *Los Angeles Times*, 29 July 1958. Clipping in KTLA archives.

[127] "Telecopter! Here's How It Works," *TV-Radio Life*, 20 September 1958. KTLA archives.

[128] "Bird's-Eye View," *Time*, undated clipping in KTLA archive, p. 51.

[129] Larry Scheer interview, 16 June 1970, unpublished transcript in KTLA archives.

[130] Gary Greenfield, "KTLA: You Ain't Heard Nothin' Yet!" Unpublished station history in KTLA archives, 24 October 1972, p. 3.

[131] Chambers, *News at Ten*, pp. 118-119.

[132] "Telecopter First On Scene of Drowning at Brando Home," Press release in KTLA archives, 12 September 1958.

[133] *Los Angeles Times*, 12 September 1958, page 1.

[134] Greenfield, op. cit.

[135] KTLA reprint from *Broadcasting* magazine, 7 December 1959. KTLA archives.

[136] KTLA press release, 24 July 1958, John Silva private collection.

[137] The account of the Bel Air fire is drawn from newspaper and magazine clippings in the KTLA archives and from a highlight video, "The Bel Air Fire," 6-7 November 1961, in the collection of the Museum of Radio and Television (MRT).

[138] Clete Roberts, interviewed on "Mobil Showcase, When Havoc Struck: The Bel Air Fire." MRT collection.

[139] Chambers, pp. 126-128.

[140] "Nixon Forced to Flee from His Rented Home," *Los Angeles Times*, 7 November 1961, p. 3.

[141] "Disaster Halts Filming of Movies, TV Shows," *Los Angeles Times*, 7 November 1961, p. 16.

[142] "Fred MacMurrays Flee, Home Damaged," *Los Angeles Times*, 7 November 1961, p. 5.

[143] Charles Denton, "Fire Coverage Was Local TV At Its Finest," *Los Angeles Examiner*, 8 November 1961. KTLA archives.

[144] These personal notes from the two newspaper editors were reproduced in KTLA press releases.

[145] "KTLA Special Release," undated, KTLA archives.

[146] "Celebrities Sift Ruins and Survey Damage to Homes," *Los Angeles Times*, 8 November 1961.

[147] "Radio-tv covered L.A. fire all the way: KTLA sets pace for round-the-clock fire watch," *Broadcasting*, clipping in KTLA archives.

[148] "For the Record," *TV Guide*, 18 November 1961. KTLA archives.

[149] KTLA sales handout, April 1962, KTLA archives.

[150] Denton, op. cit.

[151] Chambers, op. cit., pp. 145-146.

[152] Unless otherwise noted, the coverage description is from "The Baldwin Hills Disaster: Why?" 14 December 1963, videotape in the MRT collection, Los Angeles.

[153] TV Guide, quoted in Milt Valera, "Rotary-winged TV Station, Los Angeles Viewers Get their News Live and in Living Color, Via Copter," *Rotor & Wing*, Vol. 4 No. 8, August 1970, cover and pp. 19-21, 32-33.

[154] Dick Adler, "Scheer and Telecopter Winging It to KNBC," Los Angeles Times, 19 July 1974, Part IV, p. 20.

[155] Dale Fetherling, "Fatal Milieu: Tragedy May Have Stalked Gary Powers," *Los Angeles Times*, 28 August 1977, Sec. 1, pp. 3, 16 and 17.

[156] Dale Fetherling, "Wreckage of Powers Helicopter Examined," *Los Angeles Times*, 3 August 1977, Section II, page 1.

[157] "Putting television to flight: John Silva, NAB's engineer for 1974," *Broadcasting*, 31 March 1975, p. 109.

[158] Stan Chambers, *News at Ten,* p. 65.

[159] Sig Mickelson, *The Decade that Shaped Television News: CBS in the 1950s*, (Westport, CT: Praeger, 1998), p. 34.

[160] Matusow, *The Evening Stars*, p. 157.

[161] Ritchie. *Please Stand By,* p. 124.

[162] Hinds, op. cit., p. 39.

[163] Mickelson, op. cit., p. 9.

[164] *New York Times*, 2 July 1939, p. 8, quoted in Nielsen, Theodore L. *A History of Chicago Television News Presentation (1948-1968)*. University of Wisconsin, 1971. p. 39.

[165] Mickelson, op. cit., p. 10.

[166] Ritchie, op. cit., pp. 89-90.

[167] Matusow, op. cit., pp. 69-70.

[168] Ritchie, op. cit.

[169] Nielsen, op cit., p. 280.

[170] For a detailed history of print journalists in films of the 1930s and 40s see Joe Saltzman, *The Image of the Journalist in Popular Culture*, Los Angeles, University of Southern California, 2000.

[171] William C. Eddy, *Television: The Eyes of Tomorrow.* (New York: Prentice Hall Inc. 1945), p. 277. Cited in Nielsen, page 45.

[172] Ritchie, op. cit., p. 120.

[173] Chambers, op. cit., p. 47.

[174] Commercials, videotape #VA7698T in UCLA Film and TV archive.

[175] Eddy, op.cit., p. 276. Quoted in Nielsen, p. 39.

[176] Martha D. Jones, "The Most Attractive Man on Television Today," TELE-Views, 20 January 1950, p. 4. Clipping in KTLA archives.

[177] Chambers, op. cit., pp. 44-45.

[178] Landsberg, cited in Williams, *"Remote" Possibilities,* p. 136.

[179] Ritchie, op. cit., p. 123.

[180] Matusow, op.cit., introduction.

[181] William Ray, "Methods of Presenting News on Television," Baskett Mosse and Fred Whiting (eds.), *Television News Handbook*, (Evanston, Medill School of Journalism, 1953), p. 5. Cited in Nielsen, p. 201.

[182] Larry Wolters, "TV Newscasting Still has a Hard Road to Travel," *Chicago Tribune,* 3 March 1949, cited in Nielsen, p. 202.

[183] Hinds, op. cit., pp. 117-118.

[184] Nielsen, op. cit., p. 205

[185] Orrin E. Dunlap Jr. *The Future of Television,* (New York: Harper and Brothers, 1942), p. 131. Cited in Nielsen, p. 32.

[186] David Allen, "George Putnam, Still Talking Back," *Inland Valley Daily Bulletin*, 27 August 2000, p. A6.

[187] Claudia Puig, "Q & A with George Putnam: 'I Broadcast in My Spare Time,'" *Los Angeles Times*, 21 July 1994, Calendar section, p. F-1.

[188] Allen, op. cit.

[189] Allen, op. cit.

[190] Bill Stout appeared on "George Putnam: Fifty Years in Broadcasting," KTTV special, 13 July 1984. Videotape #VA11378T in UCLA archive.

[191] Sam Zelman interview, 7 April 1989, videotape in USC School of Journalism archives.

[192] Bob Voight, Letter to the Editor, Los Angeles Times, 14 June 1998, p. M-4.

[193] Stout, op. cit.

[194] Allen, op. cit.

[195] Puig, op. cit.

[196] Robert S. Alley and Irby B. Brown, *Love is All Around: The Making of the Mary Tyler Moore Show*, (New York: Dell, 1989), pp. 3-10.

[197] Allan Burns interview with author, 17 October 2000. All Burns quotes from this source unless otherwise noted.

[198] Howard Gingold, "Standards? Whose? KCBS' or KNXT's?" *Los Angeles Times*, 16 March 1992, Calendar Section, p. F-3, Counterpunch.

[199] Jerry Dunphy interview, 14 April 1989, videotape in USC School of Journalism archives. All Dunphy quotes from this source unless otherwise noted.

[200] Joe Saltzman, "Television News: Notes from a Survivor," unpublished manuscript in USC archives.

[201] Dunphy interview, op. cit.

[202] Nielsen, pp. 212-213.

[203] Matusow, pp. 156-157.

[204] Saltzman, "Notes from a Survivor," pp. 2-5.

[205] Alley and Brown, op. cit., p. 118.

[206] Puig, op. cit.

[207] Allen, op. cit.

[208] Larry Kane, *Larry Kane's Philadelphia*, (Philadelphia: Temple University Press, 2000), pp. 43-48.

[209] Clay Gowran, "Many Laud that Fahey Bow Tie," *Chicago Tribune*, 26 February 1968. Section 2, p. 23.

[210] Gerald Ruben interview with author.

[211] Bill Boyarsky, "The Question That Joined the Battle," *Los Angeles Times*, 12 April 1991, p. B-2.

[212] "George Putnam: Fifty Years in Broadcasting, op. cit.

[213] Saltzman, *The Image of the Journalist in Popular Culture*, p. 123.

[214] Alley and Brown, op.cit., p. 119.

www.ingramcontent.com/pod-product-compliance
Lightning Source LLC
Chambersburg PA
CBHW071400210526
45465CB00001B/183